Statistical Physics

Advanced Texts in Physics

This program of advanced texts covers a broad spectrum of topics that are of current and emerging interest in physics. Each book provides a comprehensive and yet accessible introduction to a field at the forefront of modern research. As such, these texts are intended for senior undergraduate and graduate students at the M.S. and Ph.D. levels; however, research scientists seeking an introduction to particular areas of physics will also benefit from the titles in this collection.

Claudine Hermann

Statistical Physics

Including Applications
to Condensed Matter

With 63 Figures

 Springer

Claudine Hermann
Laboratoire de Physique de la Matière Condensée
Ecole Polytechnique
91128 Palaiseau
France
claudine.hermann@polytechnique.edu

Library of Congress Cataloging-in-Publication Data is available.

ISBN 978-1-4419-1980-9   ISBN 978-0-387-25099-1   Printed on acid-free paper.

Printed in the United States of America.   (MPY)

9 8 7 6 5 4 3 2 1

springeronline.com

Table of Contents

Introduction

A glosssary at the end of this introduction defines the terms specific to Statistical Physics. In the text, these terms are marked by an asterisk in exponent.

A course of Quantum Mechanics, like the one taught at Ecole Polytechnique, is devoted to the description of the state of an individual particle, or possibly of a few ones. Conversely, the topic of this book will be the study of *systems**
containing very many particles, of the order of the Avogadro number $\mathcal{N}$, for example the molecules in a gas, the components of a chemical reaction, the adsorption sites for a gas on a surface, the electrons of a solid. You certainly previously studied this type of system, using Thermodynamics which is ruled by "exact" laws, such as the ideal gas one. Its physical parameters, that can be measured in experiments, are macroscopic quantities like its pressure, volume, temperature, magnetization, etc.

It is now well-known that the correct microscopic description of the state of a system, or of its evolution, requires the Quantum Mechanics approach and the solution of the Schroedinger equation, but how can this equation be solved when such a huge number of particles comes into play ? Printing on a listing the positions and velocities of the $\mathcal{N}$ molecules of a gas would take a time much longer that the one elapsed since the Big Bang ! A statistical description is the only issue, which is the more justified as the studied system is larger, since the *relative fluctuations* are then very small (Ch. 1).

This course is restricted to systems in *thermal equilibrium** : to reach such an equilibrium it is mandatory that interaction terms should be present in the hamiltonian of the total system : even if they are weak, they allow energy exchanges between the system and its environment (for example one may be concerned by the electrons of a solid, the environment being the ions of the same solid). These interactions provide the way to equilibrium. This approach takes place during a characteristic time, the so-called "relaxation time", with a range of values which depends on the considered system (the study of off-equilibrium phenomena, in particular transport phenomena, is another branch of the field of Statistical Physics, which is not discussed in this book). Whether the system is in equilibrium or not, the parameters accessible to an experiment are just those of Thermodynamics. The purpose of Statistical Physics is to bridge the gap between the microscopic modeling of the system and the macroscopic physical parameters that characterize it (Ch. 2).

The studied system can be found in a great many states, solutions of the Quantum Mechanics problem, which differ by the values of their microscopic parameters while satisfying the same specified macroscopic physical conditions, for example a fixed number of particles and a given temperature. The aim is thus to propose statistical hypotheses on the likelihood for a particular microscopic state or another one to be indeed realized in these conditions : the statistical description of systems in equilibrium is based on the postulate on the quantity *statistical entropy* : it should be *maximum* consistently with the constraints on the system under study. The treated problems concern all kinds of degrees of freedom : translation, rotation, magnetization, sites occupied by the particles, etc. For example, for a system made of N spins, of given total magnetization M, from combinatory arguments one will look for all the microscopic spin configurations leading to M ; then one will make the basic hypothesis that, in the absence of additional information, all the possible configurations are equally likely (Ch. 2). Obviously it will be necessary to verify the validity of the microscopic model, as this approach must be consistent with the laws and results of Thermodynamics (Ch. 3)

It is specified in the Quantum Mechanics courses that the limit of Classical Mechanics is justified when the de Broglie wavelength associated with the wavefunction is much shorter than all the characteristic dimensions of the problem. When dealing with free indistinguishable mobile particles, the characteristic length to be considered is the average distance between particles, which is thus compared to the de Broglie wavelength, or to the size of a typical wave packet, at the considered temperature.

According to this criterion, among the systems including a very large number of mobile particles, all described by Quantum Mechanics, two types will thus be distinguished, and this will lead to consequences on their Statistical Physics properties :

– in dilute enough systems the wave packets associated with two neighboring particles do not overlap. In such systems, the possible potential energy of a particle expresses an external field force (for example gravity) or its confinement in a finite volume. This system constitutes THE *"ideal gas"* : this means that all diluted systems of mobile particles, subjected to the same external potential, have the same properties, except for those related to the particle mass (Ch. 4) ;

– in the opposite case of a high density of particles, like the atoms of a liquid or the electrons of a solid, the wave packets of neighboring particles do overlap. Quantum Mechanics analyzes this latter situation through the *Pauli principle* : you certainly learnt a special case of it, the Pauli exclusion principle, which applies to the filling of atomic levels by electrons in Chemistry. The general expression of the Pauli principle (Ch. 5) specifies the conditions on the *symmetry* of the *wavefunction for N identical particles*. There are only two possibilities and to each of them is associated a type of particle :

- on the one hand, the *fermions*, such that only a single fermion can be in a given quantum state;
- on the other hand, the *bosons*, which can be in unlimited number in a determined quantum state.

The consequences of the Pauli principle on the statistical treatment of non-interacting indistinguishable particles, i.e., the two types of Quantum Statistics, that of Fermi-Dirac and that of Bose-Einstein, are first presented in very general terms in Ch. 6.

In the second part of this course (Ch. 7 and the following chapters), examples of systems following Quantum Statistics are treated in detail. They are very important for the physics and technology of today. Some properties of fermions are presented using the example of *electrons in metallic* (Ch. 7) or, more generally, *crystalline* (Ch. 8) solids.

Massive boson particles, in conserved number, are studied using the examples of the superfluid helium and the Bose-Einstein condensation of atoms. Finally, the *thermal radiation*, an example of a system of bosons in non-conserved number, will introduce us into very practical current problems (Ch. 9).

A topic in physics can only be fully understood after sufficient practice. A selection of exercises and problems with their solution is presented at the end of the book.

This introductory course of Statistical Physics emphasizes the microscopic interpretation of results obtained in the framework of Thermodynamics and illustrates its approach, as much as possible, through practical examples : thus the Quantum Statistics will provide an opportunity to understand what is an insulator, a metal, a semiconductor, or what is the principle of the greenhouse effect that could deeply influence our life on earth (and particularly that of our descendants !).

The content of this book is influenced by the previous courses of Statistical Physics taught at Ecole Polytechnique : the one by Roger Balian *From microphysics to macrophysics : methods and application to statistical physics*, volume I translated by D. ter Haar and J.F. Gregg, volume II translated by D.

ter Haar, Springer Verlag Berlin (1991), given during the 1980s; the course by Edouard Brézin during the 1990s. I thank them here for all that they brought to me in the stimulating field of Statistical Physics.

Several discussions in the present book are inspired from the course by F. Reif *Fundamentals of Statistical and Thermal Physics*, Mac Graw-Hill (1965), a not so recent work, but with a clear and practical approach, that should suit students attracted by the "physical" aspect of arguments. Many other Physical Statistics text books, of introductory or advanced level, are edited by Springer. This one offers the point of view of a top French scientific Grande Ecole on the subject.

The present course is the result of a collective work of the Physics Department of Ecole Polytechnique. I thank the colleagues with whom I worked the past years, L. Auvray, G. Bastard, C. Bachas, B. Duplantier, A. Georges, T. Jolicoeur, M. Mézard, D. Quéré, J-C. Tolédano, and particularly my co-workers from the course of Statistical Physics "A", F. Albenque, I. Antoniadis, U. Bockelmann, J.-M. Gérard, C. Kopper, J.-Y. Marzin, who brought their suggestions to this book and with whom it is a real pleasure to teach.

Finally, I would like to thank M. Digot, M. Maguer and D. Toustou, from the Printing Office of Ecole Polytechnique, for their expert and good-humored help in the preparation of the book.

Glossary

We begin by recalling some definitions in Thermodynamics that will be very useful in the following. For convenience, we will also list the main definitions of Statistical Physics, introduced in the next chapters of this book. This section is much inspired by chapter 1, The Language of Thermodynamics, from the book *Thermodynamique*, by J.-P. Faroux and J. Renault (Dunod, 1997).

Some definitions of Thermodynamics :

The *system* is the object under study ; it can be of microscopic or macroscopic size. It is distinguished from the rest of the Universe, called *the surroundings*.

A system is *isolated* if it does not exchange anything (in particular energy, particles) with its surroundings. The *parameters* (or *state variables*) are independent quantities which define the macroscopic state of the system : their nature can be mechanical (pressure, volume), electrical (charge, potential), thermal (temperature, entropy), etc. If these parameters take the same value at any point of the system, the system is *homogeneous*. The *external* parameters are independent parameters (volume, electrical or magnetic field) which can be controlled from the outside and imposed to the system, to the experimental accuracy. The *internal* parameters cannot be controlled and may fluctuate ; this is the case for example of the local repartition of density under an external constraint. The internal parameters adjust themselves under the effect of a modification of the external parameters.

A system is in *equilibrium* when all its internal variables remain constant in time and, in the case of a system which is not isolated, if it has no exchange with its surroundings : that is, there is no exchange of energy, of electric charges, of particles. In Thermodynamics it is assumed that any system, submitted to constant and uniform external conditions, evolves toward an equilibrium state that it can no longer spontaneously leave afterward. The thermal equilibrium between two systems is realized after exchanges between themselves : this is not possible if the walls which separate them are *adiabatical*, i.e., they do not transmit any energy.

In a homogeneous system, the *intensive* parameters, such as its temperature, pressure, the difference in electrical potential, do not vary when the system volume increases. On the other hand, the *extensive* parameters such as the volume, the internal energy, the electrical charge are proportional to the volume.

Any evolution of the system from one state to another one is called a *process* or *transformation*. An *infinitesimal* process corresponds to an infinitely small variation of the external parameters between the initial state and the final state of the system. A *reversible* transformation takes place through a continuous set of equilibrium intermediate states, for both the system and the surroundings, i.e., all the parameters defining the system state vary continuously : it is then possible to vary these parameters in the reverse direction and return to the initial state. A transformation which does not obey this definition is said to be *irreversible*.

In a *quasi-static* transformation, at any time the system is in internal equilibrium and its internal parameters are continuously defined. Contrarly to a reversible process, this does not imply anything about the surroundings but only means that the process is slow enough with respect to the characteristic relaxation time of the system.

Some definitions of Statistical Physics :

A configuration defined by the data of the microscopic physical parameters, given by Quantum or Classical Mechanics, is a *microstate*.

A configuration defined by the data of the macroscopic physical parameters, given by Thermodynamics, is a *macrostate*.

An *ensemble average* is performed at a given time on an assembly of systems of the same type, prepared in the same macroscopic conditions.

In the *microcanonical* ensemble, each of these systems is isolated and its energy E is fixed, that is, it is lying in the range between E and $E + \delta E$. It contains a fixed number N of particles.

In the *canonical* ensemble, each system is in thermal contact with a large system, a heat reservoir, which imposes its temperature T ; the energy of each system is different but for macroscopic systems the average $\langle E \rangle$ is defined with very good accuracy. Each system contains N particles.

In the canonical ensemble the *partition function* is the quantity that norms the probabilities, which are Boltzmann factors. (see Ch. 2)

In the *grand canonical* ensemble, each system is in thermal contact with a heat reservoir which imposes its temperature T, the energy of each system being different. The energies of the various systems are spread around an average value, defined with high accuracy in the case of a macroscopic system. Each system is also in contact with a particle reservoir, which imposes its chemical potential : the number of particles differs according to the system, the average value $\langle N \rangle$ is defined with high accuracy for a macroscopic system.

Chapter 1

Statistical Description of Large Systems. Postulates of Statistical Physics

The aim of Statistical Physics is to bridge the gap between the microscopic and macroscopic worlds. Its first step consists in stating hypotheses about the microscopic behavior of the particles of a macroscopic system, i.e., with characteristic dimensions very large with respect to atomic distances; the objective is then the prediction of macroscopic properties, which can be measured in experiments. The system under study may be a gas, a solid, etc., i.e., of physical or chemical or biological nature, and the measurements may deal with thermal, electrical, magnetic, chemical, properties. In the present chapter we first choose a microscopic description, either through Classical or Quantum Mechanics, of an individual particle and its degrees of freedom. The phase space is introduced, in which the time evolution of such a particle is described by a trajectory : in the quantum case, this trajectory is defined with a limited resolution because of the Heisenberg uncertainty principle (§ 1.1). Such a description is then generalized to the case of the very many particles in a macroscopic system : the complexity arising from the large number of particles will be suggested from the example of molecules in a gas (§ 1.2). For so large numbers, only a statistical description can be considered and § 1.3 presents the basic postulate of Statistical Physics and the concept of statistical entropy, as introduced by Ludwig Boltzmann (1844-1906), an Austrian physicist, at the end of the 19th century.

1.1   Classical or Quantum Evolution of a Particle ; Phase Space

For a single particle we describe the time evolution first in a classical framework, then in a quantum description. In both cases, this evolution is conveniently described in the phase space.

1.1.1   Classical Evolution

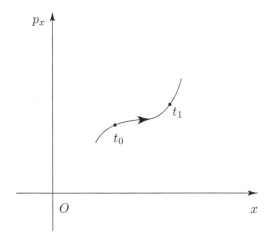

FIG. 1.1 : Trajectory of a particle in the phase space.

Consider a single classical particle, of momentum $\vec{p}_0$, located at the coordinate $\vec{r}_0$ at time t_0. It is submitted to a force $\vec{F}_0(t)$. Its time evolution can be predicted through the Fundamental Principle of Dynamics, since

$$\frac{d\vec{p}}{dt} = \vec{F}(t) \tag{1.1}$$

This evolution is deterministic since the set $(\vec{r}, \vec{p})$ can be deduced at any later time t. One introduces the one-particle "phase space", at six dimensions, of coordinates (x, y, z, p_x, p_y, p_z). The data $(\vec{r}, \vec{p})$ correspond to a given point of this space, the time evolution of the particle defines a trajectory, schematized on the Fig. 1.1 in the case of a one-dimension space motion. In the particular case of a periodic motion, this trajectory is closed since after a period the particle returns at the same position with the same momentum : for example the abscissa x and momentum p_x of a one-dimension harmonic oscillator of mass m and frequency ω are linked by :

$$\frac{p_x^2}{2m} + \frac{1}{2}m\omega^2 x^2 = E \tag{1.2}$$

In the phase space this relation is the equation of an ellipse, a closed trajectory periodically described (Fig. 1.2). Another way of determining the time

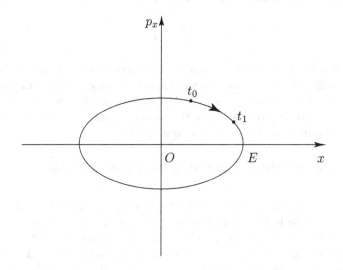

FIG. 1.2: Trajectory in the phase space for a one-dimension harmonic oscillator of energy E.

evolution of a particle is to use the Hamilton equations (see for example the course of *Quantum Mechanics* by J.-L. Basdevant and J. Dalibard, including a CDROM, Springer, 2002) : the motion is deduced from the position $\vec{r}$ and its corresponding momentum $\vec{p}$. The hamiltonian function associated with the total energy of the particle of mass m is given by :

$$h = \frac{p^2}{2m} + V(\vec{r}) \tag{1.3}$$

where the first term is the particle kinetic energy and V is its potential energy, from which the force in (1.1) is derived. The Hamilton equations of motion are given by :

$$\begin{cases} \dot{\vec{r}} = \dfrac{\partial h}{\partial \vec{p}} \\[2ex] \dot{\vec{p}} = -\dfrac{\partial h}{\partial \vec{r}} \end{cases} \tag{1.4}$$

and are equivalent to the Fundamental Relation of Dynamics (1.1).

Note : Statistical Physics also applies to relativistic particles which have a different energy expression.

1.1.2   Quantum Evolution

It is well known that the correct description of a particle and of its evolution requires Quantum Mechanics and that to the classical hamiltonian function h corresponds the quantum hamiltonian operator $\hat{h}$. The Schroedinger equation provides the time evolution of the state $|\psi\rangle$:

$$i\hbar\frac{\partial|\psi\rangle}{\partial t} = \hat{h}|\psi\rangle \tag{1.5}$$

The squared modulus of the spatial wave function associated with $|\psi\rangle$ gives the probability of the location of the particle at any position and any time. Both in Classical and Quantum Mechanics, it is possible to reverse the time direction, i.e., to replace t by $-t$ in the motion equations (1.1), (1.4) or (1.5), and thus obtain an equally acceptable solution.

1.1.3   Uncertainty Principle and Phase Space

You know that Quantum Mechanics introduces a probabilistic character, even when the particle is in a well-defined state : if the particle is not in an eigenstate of the measured observable $\hat{A}$, the measurement result is uncertain. Indeed, for a *single measurement* the result is any of the eigenvalues a_α of the considered observable. On the other hand, when this measurement is *reproduced* a large number of times on identical, similarly prepared systems, the average of the results is $\langle\psi|\hat{A}|\psi\rangle$, a well-defined value.

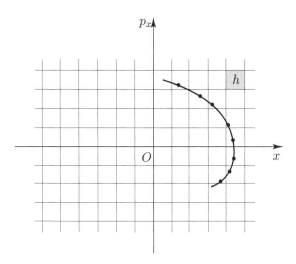

FIG. 1.3: A phase space cell, of area h for a one-dimension motion, corresponds to one quantum state.

In the phase space, this quantum particle has no well-defined trajectory since, at a given time t, in space and momentum coordinates there is no longer an exact position, but a typical extent of the wave function inside which this particle can be found.

Another way of expressing this difficulty is to state the Heisenberg uncertainty principle : if the particle abscissa x is known to Δx, then the corresponding component p_x of its momentum cannot be determined with an accuracy better than Δp_x, such that $\Delta x \cdot \Delta p_x \geq \hbar/2$, where $\hbar = h/2\pi$, h being the Planck constant. The "accuracy" on the phase space trajectory is thus limited, in other words this trajectory is "blurred" : the phase space is therefore divided into cells of area of the order of h for a one-dimension motion, h^3 if the problem is in three dimensions, and the state of the considered particle cannot be defined better than within such a cell (Fig. 1.3). This produces a discretization of the phase space. (In the case of a classical particle, it is not *a priori* obvious that such a division is necessary, and if it is the case, that the same cell area should be taken. It will be stated in § 1.2.2 that, for sake of consistency at the classical limit of the quantum treatment, this very area should be chosen.)

1.1.4   Other Degrees of Freedom

Up to now we have only considered the degrees of freedom related to the particle position, which take continuous values in Classical Mechanics and are described by the spatial part of the wave function in Quantum Mechanics. If the particle carries a magnetic moment, the wave function includes an additional component, with discrete eigenvalues (see a Quantum Mechanics course).

1.2   Classical Probability Density ; Quantum Density Operator

When considering a macroscopic system, owing to the extremely large number of parameters, the description of the particles' evolution can only be done through a statistical approach. The tools are different for either the classical or the quantum description of the particles' motion.

1.2.1   Necessity of a Statistical Approach for Macroscopic Systems

The physical parameters referring to individual particles, the evolution of which was described in the above section, are measured at the atomic scale (distances of the order of a Bohr radius of the hydrogen atom, i.e., a tenth of a nm, energies of the order of a Rydberg or of an electron-volt, characteristic times in the range of the time between two collisions in a gas, that is about 10^{-10} sec for N_2 in standard conditions). For an object studied in Thermodynamics or measured in usual laboratory conditions, the scales are quite different : the dimensions are currently of the order of a mm or a cm, the energies are measured in joules, the measurement time is of the order of a second. Thus two very different ranges are evidenced : the microscopic one, described by Classical or Quantum Mechanics laws, unchanged when reversing the direction of time ; the macroscopic range, which is often the domain of irreversible phenomena, as you learnt in previous courses. Now we will see what are the implications when shifting from the microscopic scale to the macroscopic one. For that we will first consider the case of a few particles, then of a number of particles of the order of the Avogadro number.

If the physical system under study contains only a few particles, for example an atom with several electrons, the behavior of each particle follows the Mechanics laws given above. Once the position and momentum of each particle are known at $t = t_0$, together with the hamiltonian determining the system evolution, in principle one can deduce $\vec{r}$ and $\vec{p}$ at any further time t (yet the calculations, which are the task of Quantum Chemistry in particular, quickly become very complex).

Thus in principle the problem can be solved, but one is facing practical difficulties, which quickly become "impossibilities", as soon as the system under study contains "many" particles. Indeed in Molecular Dynamics, a branch of Physics, the evolution of particle assemblies is calculated according to the Classical Mechanics laws, using extremely powerful computers. However, considering the computing capacities available today, the size of the sample is restricted to several thousands of particles at most.

Consider now a macroscopic situation, like those described by Thermodynamics : take a monoatomic gas of $\mathcal{N}$ molecules, where $\mathcal{N} = 6.02 \times 10^{23}$ is the Avogadro number, and assume that at time t_0 the set of data $(\vec{r}_0, \vec{p}_0)$ is known for each particle. Without prejudging the time required for their calculation, just printing the coordinates of all these molecules at a later time t, with a printer delivering the coordinates of one molecule per second, would take 2×10^{16} years! The information issued in one day would be for 10^5 molecules only. Another way to understand the enormity of such an information on $\mathcal{N}$ molecules is to wonder how many volumes would be needed to print it.

The answer is : many more than the total of books manufactured since the discovery of printing!

One thus understands the impossibility of knowing the position and momentum of each particle at any time. In fact, the measurable physical quantities (for example volume, pressure, temperature, magnetization) only rarely concern the properties of an individual molecule at a precise time : if the recent achievements of near-field optical microscopy or atomic-force microscopy indeed allow the observation of a single molecule, and only in very specific and favorable cases, the duration of such experiments is on a "human" scale (in the range of a second). One almost always deals with physical quantities concerning an assembly of molecules (or particles) during the observation time.

"It is sufficient for the farmer to be certain that a cloud burst and watered the ground and of no use to know the way each distinct drop fell. Take another example : everybody understands the meaning of the word "granite", even if the shape, chemical composition of the different crystallites, their composition ratios and their colors are not exactly known. We thus always use concepts which deal with the behavior of large scale phenomena without considering the isolated processes at the corpuscular scale." (W. Heisenberg, "Nature in the contemporary physical science, " from the French translation, p.42 Idées NRF, Paris (1962)).

Besides, some of the observed macroscopic properties may have no microscopic equivalent :

"Gibbs was the first one to introduce a physical concept which can apply to a natural object only if our knowledge of this object is incomplete. For example if the motions and positions of all the molecules in a gas were known, speaking of the temperature of this gas would no longer retain a meaning. The temperature concept can only be used when a system is insufficiently determined and statistical conclusions are to be deduced from this incomplete knowledge." (W. Heisenberg, same reference, p.45).

It is impossible to collect the detailed information on all the molecules of a gas, anyway all these data would not really be useful. A more global knowledge yielding the relevant physical parameters is all that matters : a statistical approach is thus justified from a physical point of view. Statistics are directly related to probabilities, i.e., a large number of experiments have to be repeated either in time or on similar systems (§ 1.2.2). Consequently, one will have to consider an assembly (or an "ensemble") consisting of a large number of systems, prepared in a similar way, in which the microscopic structures differ and the measured values of the macroscopic parameters are not necessarily identical : the probability of occurrence of a particular "event" or a particular measurement is equal to the fraction of the ensemble systems for which it occurs (see §1.3.2).

1.2.2   Classical Probability Density

The hamiltonian is now a function of the positions and momenta of the N particles, i.e., $H(\vec{r}_1, \dots \vec{r}_i, \dots \vec{r}_N, \vec{p}_1, \dots \vec{p}_i, \dots \vec{p}_N)$.

Using a description analogous to that of the one-particle phase space of §1.1.1, the system of N classical particles now has a trajectory in a new phase space, at $6N$ dimensions since there are $3N$ space and $3N$ momenta coordinates. As it is impossible to exactly know the trajectory, owing to the lack of information on the coordinates of the individual particles (see §1.2.1), one can only speak of the probability of finding the system particles at time t in the neighborhood of a given point of the phase space, of coordinates $(\vec{r}_1, \dots \vec{r}_i, \dots \vec{r}_N, \vec{p}_1, \dots \vec{p}_i, \dots \vec{p}_N)$ to $\prod_{i=1}^{N} d^3\vec{r}_i d^3\vec{p}_i$.

On this phase space, a *measure* $d\tau$ and a *probability density* D have to be defined. Then the probability will be given by $D(\vec{r}_1, \dots, \vec{r}_i, \dots, \vec{p}_1, \dots, \vec{p}_i, \dots t)d\tau$.

From the Liouville theorem (demonstrated in Appendix 1.1) the *phase space elementary volume* $\prod_{i=1}^{N} d^3\vec{r}_i d^3\vec{p}_i$ is conserved for any time t during the particles evolution. A consequence is that for a macroscopic equilibrium, the only situation considered in the present book, the probability density is time-independent.

The elementary volume of the N-particle phase space is homogeneous to an action to the power $3N$, that is, $[\text{mass} \times (\text{length})^2 \times (\text{time})^{-1}]^{3N}$, and thus has the same dimension as h^{3N}, the Planck constant to the same power. The infinitesimal volume will always be chosen large with respect to the volume of the elementary cell of the N-particle phase space, so that it will contain a large number of such cells. (The volume of the elementary cell will be chosen as h^{3N}, from arguments similar to the ones for a single particle.) Finally, $d\tau$ may depend on N, the particle number, through a constant C_N.

Here we will take $d\tau = \dfrac{C_N}{h^{3N}} \prod_{i=1}^{N} d^3\vec{r}_i d^3\vec{p}_i$. The quantity $d\tau$ is thus dimensionless and is related to the "number of states" inside the volume $d^3\vec{r}_i d^3\vec{p}_i$; the constant C_N is determined in §4.4.3 and 6.7 so that $d\tau$ should indeed give the *number of states in this volume* : this choice has a interpretation in the quantum description, and the continuity between classical and quantum descriptions is achieved through this choice of C_N.

The classical *probability density* $D(\vec{r}_1, \ldots \vec{r}_i, \ldots, \vec{p}_1, \ldots, \vec{p}_i, \ldots, t)$ is then defined : it is such that $D(\vec{r}_1, \ldots \vec{r}_i, \ldots, \vec{p}_1, \ldots, \vec{p}_i, \ldots, t) \geq 0$, and

$$\int D(\vec{r}_1, \ldots \vec{r}_i, \ldots, \vec{p}_1, \ldots \vec{p}_i, \ldots, t)d\tau = 1 \tag{1.6}$$

on the whole phase space.

The *average value* at a given time t of a physical parameter $A(\vec{r}_i, \vec{p}_i, t)$ is then calculated using :

$$\langle A(t) \rangle = \int A(\vec{r}_i, \vec{p}_i, t)D(\vec{r}_i, \vec{p}_i, t)d\tau \tag{1.7}$$

In the same way, the standard deviation σ_A, representing the fluctuation of the parameter A around its average value at a given t, is calculated from :

$$(\sigma_A(t))^2 = \langle A^2(t) \rangle - \langle A(t) \rangle^2 = \int \left[A(\vec{r}_i, \vec{p}_i, t) - \langle A(\vec{r}_i, \vec{p}_i, t) \rangle \right]^2 D(\vec{r}_i, \vec{p}_i, t)d\tau \tag{1.8}$$

1.2.3 Density Operator in Quantum Mechanics

By analogy with the argument of § 1.1.2, if at time t_0 the total system is in the state $|\psi\rangle$, now with N particles, the Schroedinger equation allows one to deduce its evolution through :

$$i\hbar \frac{\partial |\psi\rangle}{\partial t} = \hat{H}|\psi\rangle \tag{1.9}$$

where $\hat{H}$ is the hamiltonian of the total system with N particles.

It was already pointed out that, when the system is in a *certain* state $|\psi\rangle$, the result of a *single measurement* of the observable $\hat{A}$ on this system is uncertain (except when $|\psi\rangle$ is an eigenvector of $\hat{A}$) : the result is one of the eigenvalues a_α of $\hat{A}$, with the probability $|\langle \psi | \varphi_\alpha \rangle|^2$, associated with the projection of $|\psi\rangle$ on the eigenvector $|\varphi_\alpha\rangle$ corresponding to the eigenvalue a_α (that we take here to be nondegenerate, for sake of simplicity). If the measurement is repeated many times, the average of the results is $\langle \psi | \hat{A} | \psi \rangle$.

Now another origin of uncertainty, of a different nature, must also be included if the state of the system at time t is already *uncertain*. Assume that the macroscopic system may be found in different orthonormal states $|\psi_n\rangle$, with probabilities p_n $(0 \leq p_n \leq 1)$. In such a case the results average for repeated measurements will be :

$$\langle \hat{A} \rangle = \sum_n p_n \langle \psi_n | \hat{A} | \psi_n \rangle = \sum_n p_n \langle \psi_n | \psi_n \rangle \langle \psi_n | \hat{A} | \psi_n \rangle = \mathrm{Tr}\,(\hat{D}\hat{A}) \qquad (1.10)$$

The density operator, defined by

$$\hat{D} = \sum_n p_n |\psi_n\rangle \langle \psi_n| \qquad (1.11)$$

has been introduced (its properties are given in Appendix 1.2). This operator includes both our imperfect knowledge of the system's quantum state, through the p_n factors, and the fluctuations related to the quantum measurement, which would already be present if the state $|\psi_n\rangle$ was certain (this state appears through its projector $|\psi_n\rangle\langle\psi_n|$). One verifies that

$$\mathrm{Tr}\,(\hat{D}\hat{A}) = \sum_{n'} \sum_n p_n \langle \psi_{n'} | \psi_n \rangle \langle \psi_n | \hat{A} | \psi_{n'} \rangle = \langle \hat{A} \rangle \qquad (1.12)$$

since $\langle \psi_{n'} | \psi_n \rangle = \delta_{n'n}$.

In the current situations described in the present course, the projection states $|\psi_n\rangle$ will be the eigenstates of the hamiltonian $\hat{H}$ of the N-particle total system, the eigenvalues of which are the *accessible energies* of the system. In the problems considered the operator $\hat{D}$ will thus be diagonal on such a basis : each energy eigenstate $|\psi_n\rangle$ will simply be weighted by its probability p_n.

Now hypotheses are required to obtain an expression of $D(\vec{r}_i, \vec{p}_i, t)$ in the classical description, or of the probabilities p_n in the quantum one.

1.3 Statistical Postulates ; Equiprobability

We should now be convinced that a detailed microscopic treatment is impossible for a system with a very large number of particles : we will have to limit ourselves to a statistical approach. It is necessary to define the microscopic and macroscopic states to which this description will apply.

1.3.1 Microstate, Macrostate

A configuration defined by the data of microscopic physical parameters (for example, positions and momenta of all the particles ; quantum numbers characterizing the state of each particle ; magnetization of each paramagnetic atom localized in a solid) is a *microstate**.[1] A configuration defined by the value of macroscopic physical parameters (total energy, pressure, temperature, total magnetization, etc.) is a *macrostate**. Obviously a macrostate is almost always realized by a very large number of microstates : for example, in a paramagnetic solid, permuting the values of the magnetic moments of two fixed atoms creates a new microstate associated with the same macrostate. Assume that each magnetic moment can only take two values : $+\vec{\mu}$ and $-\vec{\mu}$. The number of microstates $W(p)$ all corresponding to the macrostate of magnetization $\vec{M} = (2p - N)\vec{\mu}$, in which p magnetic moments $\vec{\mu}$ are in a given direction and $N - p$ in the opposite direction, is the number of choices of the p moments to be reversed among the total of N. It is thus equal to $C_N^p = \dfrac{N!}{p!(N - p)!}$. Since N is macroscopic and in general p too, C_N^p is a very large number.

The scope of this book is the analysis of systems in *equilibrium** : this means that one has waited long enough so that macroscopic physical parameters now keep the same values in time within very small fluctuations (the macrostate is fixed), whereas microscopic quantities may continue their evolution. Besides, even when the system is not *isolated** from its surroundings, if it has reached equilibrium, by definition there is no algebraical transport out of the system of matter, energy, etc.

The approach toward equilibrium is achieved through interactions, bringing very small amounts of energy into play, yet essential to thermalization : in a solid, for example, electrons reach an equilibrium, among themselves and with the ions, through the collisions they suffer, for example, on ions vibrating at non-vanishing temperature, or impurities. Here the characteristic time is very short, of the order of 10^{-14} sec in copper at room temperature. For other processes this time may be much longer : the time for a drink to reach its equilibrium temperature in a refrigerator is of the order of an hour. If the system has no measurable evolution during the experiment, it is considered to be in equilibrium.

1.3.2 Time Average and Ensemble Average

Once equilibrium is reached, the time average of a physical parameter, which is likely to fluctuate in a macrostate, is calculated using the classical probability density (1.7) or the quantum density operator (1.11), that is, for example

---

[1]The words marked by an asterisk are defined in the glossary preceding Chapter 1.

$$\langle A \rangle = \frac{1}{T} \int_{t'}^{t'+T} dt \int A(\vec{r}_i, \vec{p}_i, t) D(\vec{r}_i, \vec{p}_i, t) d\tau, \qquad (1.13)$$

Would the same result be obtained by calculating the average at a given time on an assembly of macrostates prepared in a similar way? A special case, allowing this issue to be understood, is that of an isolated system, of constant total energy E. In the N-particle phase space, the condition of constant energy defines a $6N - 1$-dimension surface. The points representing the various systems prepared at the same energy E are on this surface. Is the whole surface described? What is the time required for that? This equivalence of the time average with the average on an ensemble of systems prepared in the same way is the object of the ergodic theorem, almost always applicable and that we will assume to be valid in all the cases treated in this course.

In Statistical Physics, the common practice since Josiah Willard Gibbs (1839-1903) is indeed to take the average, at a fixed time, on an assembly (an ensemble) of systems of the same nature, prepared under the same macroscopic conditions : this is the so-called *ensemble average*. As an example, consider a system, replicated a large number of times, which is exchanging energy with a reservoir. The energy distribution between the system and the reservoir differs according to the replica whereas, as will be shown in § 2.3.1, the system temperature is always that of the heat reservoir. Let $\mathbf{N}$ be the number of prepared identical systems, where $\mathbf{N}$ is a very large number. We consider a physical parameter f that can fluctuate, in the present case the energy of each system. Among the $\mathbf{N}$ systems, this parameter takes the value f_l in $\mathbf{N}_l$ of them; consequently, the ensemble average of the parameter f will be equal to

$$\langle f_N \rangle = \frac{1}{\mathbf{N}} \sum_l \mathbf{N}_l f_l \qquad (1.14)$$

When $\mathbf{N}$ tends to infinity, $\mathbf{N}_l/\mathbf{N}$ tends to the probability p_l to achieve the value f_l and the ensemble average tends to the average value calculated using these probabilities :

$$\lim_{N \to \infty} \langle f_N \rangle = \langle f \rangle = \sum_l p_l f_l \qquad (1.15)$$

1.3.3   Equiprobability

On daily experience (!), when throwing a die, in the absence of any other information, it is commonly assumed that every face is equally likely to occur, so that the probability of occurrence of a given face (1 for example) is $1/6$.

In Statistical Physics, an analogous postulate is stated : one assumes that *in the absence of additional information, for an isolated system of energy ranging between E and $E + \delta E$, where δE is the uncertainty, all the accessible microstates are equally probable.* (In a classical description, this is equivalent to saying that all the accessible phase space cells are equiprobable.) Considering the example of the magnetic moments, this means that, to obtain the same magnetization value, it is equally likely to have a localized magnetic moment reversed on a particular site or on another one.

The larger the number of accessible states, the larger the uncertainty or disorder, the smaller the information about the system : a measure of the disorder is the number of accessible states, or the size of the accessible volume, of the phase space.

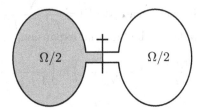

Fig. 1.4 : Joule-Gay-Lussac expansion.

Consider for example a container of volume Ω, split into two chambers of volume $\Omega/2$ each, isolated from its surroundings (Fig. 1.4). If the gas at equilibrium is completely located inside the left chamber, the disorder is smaller than if, at the same conditions, the equilibrium occurs in the total volume Ω. Opening the partition between the two chambers, in the so-called Joule-Gay-Lussac free expansion, means relaxing the constraint on the molecules to be located in the left chamber. This is an *irreversible** process[2] as the probability for a spontaneous return, at a later time, to the situation where the N molecules all lie in the left chamber is extremely low (2^{-N}). It appears that a macroscopic initial state distinct from equilibrium will most probably evolve toward a larger disorder, although the fluctuations in equilibrium are identical when the direction of time is reversed. The direction of time defined in Statistical Physics is based on the idea that a statistically improbable state was built through an external action, requiring the expending of energy.

The *basic postulate* of Statistical Physics, stated by Ludwig Boltzmann in 1877 for an *isolated system*, consists in defining a macroscopic quantity from the number of accessible microstates. This is the statistical entropy S, given

---

[2]A very interesting discussion on the character of the time direction and on irreversibility in Statistical Physics can be found in the paper by V. Ambegaokar and A.A. Clerk, American Journal of Physics, vol 67, p. 1068 (1999).

by

$$S = k_B \ln W(E) \qquad (1.16)$$

Here $W(E)$ is the number of microstates with an energy value between E and $E+\delta E$; the constant k_B is *a priori* arbitrary. To make this statistical definition of the entropy coincide with its thermodynamical expression (see § 3.3), one should take k_B equal to the Boltzmann constant, which is the ratio of the ideal gas constant $\mathcal{R} = 8.31$ J/K to the Avogadro number $\mathcal{N} = 6.02 \times 10^{23}$, that is, $k_B = \mathcal{R}/\mathcal{N} = 1.38 \times 10^{-23}$ J/K. This is what will be taken in this course. Take the example of the localized magnetic moments of a solid in the presence of an external magnetic field $\vec{B}$: in the configuration where p moments are parallel to $\vec{B}$ and $N - p$ moments are in the opposite direction, the magnetization is $\vec{M} = (2p - N)\vec{\mu}$ and the magnetic energy $E = -\vec{M} \cdot \vec{B} = (N - 2p)\vec{\mu} \cdot \vec{B}$. It was shown in § 1.3.1 that in such a case $W(E) = \dfrac{N!}{p!(N - p)!}$. For a large system, using the Stirling formula for the factorial expansion, one gets

$$S = k_B \ln W(E) = \frac{k_B N}{2} \left[\left(1 + \frac{E}{N\mu_B B} \right) \ln \left(\frac{2}{1 + (E/N\mu_B B)} \right) \right.$$
$$\left. + \left(1 - \frac{E}{N\mu_B B} \right) \ln \left(\frac{2}{1 - (E/N\mu_B B)} \right) \right] \qquad (1.17)$$

When two independent, or weakly coupled, subsystems are associated, with respective numbers of microstates W_1 and W_2, the number of microstates of the combined system is $W = W_1 \cdot W_2$, since any microstate of subsystem 1 can be associated with any microstate of 2. Consequently, $S = S_1 + S_2$, so that S is *additive**. A special case is obtained when the two subsystems are identical, then $S = 2S_1$, that is, S is *extensive**.

Formula (1.16) is fundamental in that it bridges the microscopic world (W) and the macroscopic one (S). It will be shown in § 3.3 that S is indeed identical to the entropy defined in Thermodynamics.

Before stating the general properties of the statistical entropy S as defined by (1.16), we first discuss those of $W(E)$. The aim is to enumerate the microstates of given energy E, to the measurement accuracy δE. The number of accessible states is related to both the number of particles and to the number of degrees of freedom of each particle. It is equal to the accessible volume of the phase space, to the factor C_N/h^{3N} (§ 1.2.2).

The special case of free particles, in motion in a macroscopic volume, is treated in Appendix 1.3. This example of macroscopic system allows one to understand : *i)* the extremely fast variation of $W(E)$ with E, as a power of the particle number N, assumed to be very large (a similar fast variation is also obtained for an assembly of N particles subjected to a potential); *ii)* the totally negligible effect of δE in practical cases; *iii)* the effective proportionality of $\ln W(E)$, and thus of S, to N.

In the general case, the particles may have other degrees of freedom (in the example of diatomic molecules : rotation, vibration), they may be subjected to a potential energy, etc. Results *i), ii), iii)* are still valid.

1.4 General Properties of the Statistical Entropy

As just explained on a special case, the physical parameter that contains the statistical character of the system under study is the statistical entropy. We now define it on more general grounds and relate it to the information on the system.

1.4.1 The Boltzmann Definition

The expression (1.16) of S given by Boltzmann relies on the postulate of equal probability of the W accessible microstates of an isolated system. S increases as the logarithm of the microstates number.

1.4.2 The Gibbs Definition

A more general definition of the statistical entropy was proposed by Gibbs :

$$S = -k_B \sum_i p_i \ln p_i \qquad (1.18)$$

This definition applies to an assembly of systems constructed under the same macroscopic conditions, made up in a similar way according to the process defined in § 1.3.2; in the ensemble of all the reproduced systems, p_i is the probability to attain the particular state i of the system, that is, the microstate i. It is this Gibbs definition of entropy that will be used in the remaining part of this course. For example, for systems having reached thermal equilibrium at temperature T, we will see that the probability of realizing a microstate i is higher, the lower the microstate energy (p_i is proportional to the Boltzmann factor, introduced in § 2.4.1). The Boltzmann expression (1.16) of the entropy is a special case, which corresponds to the situation where all W microstates are equiprobable, so that the probability of realizing each of them is $p_i = 1/W$.

Then :

$$S = -k_B \ln \frac{1}{W}(\sum_i p_i) = k_B \ln W \qquad (1.19)$$

It is not possible to proceed the other way and to deduce the general Gibbs definition from the particular Boltzmann definition. Yet let us try to relate the Gibbs and the Boltzmann definitions using a particular example : we are considering a system made up of $n_1 + n_2$ equiprobable microstates regrouped into two subsystems 1 and 2. Subsystem 1 contains n_1 microstates, its probability of realization is $p_1 = n_1/(n_1 + n_2)$. In the same way, subsystem 2 contains n_2 microstates, its probability of realization is equal to $p_2 = n_2/(n_1 + n_2)$.

From the Boltzmann formulation, the statistical entropy of the total system is

$$S_{\text{tot}} = k_B \ln(n_1 + n_2) \qquad (1.20)$$

This entropy is related to the absence of information on the microstate of the total system in fact realized :

– we do not know whether the considered microstate belongs subsystem 1 or 2 : this corresponds to the entropy S_{12}, which is the Gibbs term that we want to find in this example;

– a part S_1 of the entropy expresses the uncertainty inside system 1 owing to its division into n_1 microstates

$$S_1 = p_1(k_B \ln n_1) \qquad (1.21)$$

and is weighted by the probability p_1.

In the same way S_2 comes from the uncertainty inside system 2

$$S_2 = p_2(k_B \ln n_2) \qquad (1.22)$$

Since the various probabilities of realization are multiplicative, the corresponding entropies, which are related to the probabilities through logarithms, are additive :

$$S_{\text{tot}} = S_1 + S_2 + S_{12} \qquad (1.23)$$

$$k_B \ln(n_1 + n_2) = \frac{n_1}{n_1 + n_2} k_B \ln n_1 + \frac{n_2}{n_1 + n_2} k_B \ln n_2 + S_{12}$$

Then the S_{12} term is given by

$$
\begin{aligned}
S_{12} &= k_B \left(\frac{n_1 + n_2}{n_1 + n_2} \right) \ln(n_1 + n_2) \\
&\quad - k_B \left(\frac{n_1}{n_1 + n_2} \right) \ln n_1 - k_B \left(\frac{n_2}{n_1 + n_2} \right) \ln n_2 \\
&= -k_B \frac{n_1}{n_1 + n_2} \ln \frac{n_1}{n_1 + n_2} - k_B \frac{n_2}{n_1 + n_2} \ln \frac{n_2}{n_1 + n_2} \\
&= -k_B (p_1 \ln p_1 + p_2 \ln p_2)
\end{aligned}
\tag{1.24}
$$

Indeed this expression is the application of the Gibbs formula to this particular case.

On the other hand, expression (1.16) of the Boltzmann entropy is a special case of S_{Gibbs}, which makes (1.18) maximum : the information on the system is the smallest when all W microstates are equiprobable and $p_i = \dfrac{1}{W}$ for each i. The Boltzmann expression is the one that brings S to its maximum with the constraint of the fixed value of the system energy.

The expression (1.18) of S, as proposed by Gibbs, satisfies the following properties :

$S \geq 0$, for $0 \leq p_i \leq 1$ and $\ln p_i \leq 0$

$S = 0$ if the system state is certain : then $p = 1$ for this particular state, 0 for the other ones.

S is maximum when all p_i's are equal.

S is additive : indeed, when two systems 1 and 2 are weakly coupled, so that

$$
\begin{cases}
S_1 &= -k_B \sum_i p_{1i} \ln p_{1i} \\
S_2 &= -k_B \sum_j p_{2j} \ln p_{2j}
\end{cases}
\tag{1.25}
$$

by definition, the entropy of the coupled system will be written :

$$
S_{\text{tot}} = -k_B \sum_l p_l \ln p_l
\tag{1.26}
$$

Now, since the coupling is weak, each state of the total system is characterized by two particular states of subsystems 1 and 2 and $p_l = p_{1i}p_{2j}$. Consequently,

$$S_{\text{tot}} = -k_B \sum_{i,j} p_{1i}p_{2j}(\ln p_{1i} + \ln p_{2j})$$

$$= -k_B \Big[\sum_i p_{1i} \ln p_{1i} \Big(\sum_j p_{2j} \Big) + \sum_j p_{2j} \ln p_{2j} \Big(\sum_i p_{1i} \Big) \Big] \qquad (1.27)$$

$$= S_1 + S_2$$

One thus finds the additivity of S, its extensivity being a special case that is obtained by combining two systems of the same density, but of different volumes.

1.4.3 The Shannon Definition of Information

We just saw that the more uncertain the system state, or the larger the number of accessible microstates, the larger S; in addition, S is zero if the system state is exactly known. The value of S is thus related to the lack of information on the system state.

The quantitative definition of information, given by Claude Shannon (1949), is closely copied from the Gibbs definition of entropy.[3] It analyzes the capacity of communication channels. One assumes that there are W possible distinct messages, that $s_i\,(i = 0, 1, \ldots W - 1)$ is the content of the message number i, and that the probability for s_i to be emitted is p_i. Then the information content I per sent message is

$$I = -\sum_{i=0}^{W-1} p_i \log_2 p_i \qquad (1.28)$$

This definition is, to the constant $k_B\ln2$, the same as that of the Gibbs entropy, since $\log_2 p_i = \ln p_i/\ln2$.

The entropy is also related to the lack of information : probabilities are attributed to the various microstates, since we do not exactly know in which of them is the studied system. These probabilities should be chosen in such a way that they do not include unjustified hypotheses, that is, only the known properties of the system are introduced, and the entropy (the missing information) is maximized *with the constraints imposed by the physical conditions of the studied system*. This is the approach that will be chosen in the arguments of § 2.3, 2.4, and 2.5.

---

[3]See the paper by J. Machta in American Journal of Physics, vol 67, p.1074 (1999), which compares entropy, information, and algorithmics using the example of meteorological data.

Summary of Chapter 1

The time evolution of an *individual particle* is associated with a trajectory in the six-dimensional *phase space*, of coordinates $(\vec{r}, \vec{p})$. The Heisenberg uncertainty principle and the classical limit of quantum properties require that a quantum state occupies a cell of area h^3 in this space. At a given time t an *N-particle state* is represented by a point in the $6N$-dimensional phase space.

The classical statistical description of a macroscopic system utilizes a *probability density*, such that the probability of finding the system of N particles in the neighborhood of the point $(\vec{r}_1, \ldots, \vec{r}_N, \vec{p}_1, \ldots, \vec{p}_N)$, to $\prod_{i=1}^{N} d^3\vec{r}_i d^3\vec{p}_i$, is equal to

$$D(\vec{r}_1, \ldots, \vec{r}_N, \vec{p}_1, \ldots, \vec{p}_N) \frac{C_N}{h^{3N}} \prod_{i=1}^{N} d^3\vec{r}_i d\vec{p}_i$$

where C_N is a constant depending on N only.

In Quantum Mechanics the *density operator*

$$\hat{D} = \sum_n p_n |\psi_n\rangle\langle\psi_n|$$

is introduced, which contains the uncertainties related both to the incomplete knowledge of the system and to the quantum measurement.

A configuration defined by the data of the microscopic physical parameters is a *microstate*. A *macrostate* is defined by the value of macroscopic physical parameters; it is generally produced by a very large number of microstates.

This course will only deal with Statistical Physics in *equilibrium*.

The time average of a fluctuating physical parameter is generally equivalent to the average on an assembly of identical systems prepared in the same way ("ensemble average").

One assumes that, in the absence of additional information, in the case of an *isolated* system of energy between E and $E + \delta E$, where δE is the uncertainty, all the $W(E)$ accessible microstates are equally probable. The statistical entropy S is then defined by

$$S = k_B \ln W(E)$$

where $k_B = \mathcal{R}/\mathcal{N}$ is the Boltzmann constant, with $\mathcal{R}$ the ideal gas constant and $\mathcal{N}$ the Avogadro number.

The *general definition of the statistical entropy*, valid for an ensemble average, is

$$S = -k_B \sum_i p_i \ln p_i$$

where p_i is the probability of occurrence of the particular microstate i of the system. The expression for an isolated system is a special case of the latter definition. The statistical entropy is an extensive parameter, which is equal to zero when the microscopic state of the system is perfectly known.

Appendix 1.1

The Liouville Theorem in Classical Mechanics

There are two possible ways to consider the problem of the evolution between the times t and $t + dt$ of an ensemble of systems taking part in the ensemble average, each of them containing N particles, associated with the same macrostate and prepared in the same way. To *one system*, in its particular microstate, corresponds *one point* in the $6N$-dimensional phase space. Either we may be concerned with the *time evolution* of systems contained in a small volume *around the considered point* (§ A); or we may *evaluate* the number of systems that enter or leave a *fixed infinitesimal volume* of the phase space during a time interval (§ B). In both approaches it will be shown that the elementary volume of the phase space is conserved between the times t and $t + dt$.

A. Time evolution of an ensemble of systems

At time t the systems occupy the phase space volume $d\tau$ around the point with coordinates $(q_1, \ldots q_i, \ldots, q_{3N}, p_1, \ldots p_i, \ldots, p_{3N})$ (Fig. 1.5). The infinitesimal volume has the expression $d\tau = C_N \prod_{i=1}^{3N} dp_i dq_i$, where C_N is a constant that only depends on N.

To compare this volume to the one at the instant $t + dt$, one has to evaluate the products $\prod_{i=1}^{3N} dp_i(t) dq_i(t)$ and $\prod_{i=1}^{3N} dp_i(t + dt) dq_i(t + dt)$, that is, to calculate the Jacobian of the transformation

$$
\begin{cases}
p_i(t) \to p_i(t + dt) = p_i(t) + dt \cdot \dfrac{\partial p_i(t)}{\partial t} \\[2mm]
q_i(t) \to q_i(t + dt) = q_i(t) + dt \cdot \dfrac{\partial q_i(t)}{\partial t}
\end{cases}
\tag{1.29}
$$

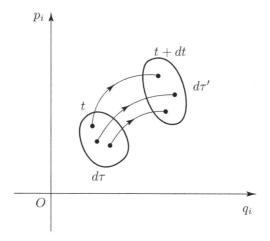

FIG. 1.5: The different systems, which were located in the volume $d\tau$ of the $6N$-dimensional phase space at t, are in $d\tau'$ at $t + dt$.

i.e., the determinant of the $6N \times 6N$ matrix

$$\begin{bmatrix} \dfrac{\partial q_i(t + dt)}{\partial q_j(t)} & \dfrac{\partial p_i(t + dt)}{\partial q_j(t)} \\[3mm] \dfrac{\partial q_i(t + dt)}{\partial p_j(t)} & \dfrac{\partial p_i(t + dt)}{\partial p_j(t)} \end{bmatrix} \tag{1.30}$$

The Hamilton equations are used, which relate the conjugated variables q_i and p_i through the N-particle hamiltonian and are the generalization of Eq. (1.4) :

$$\begin{cases} \dfrac{\partial q_i}{\partial t} = \dfrac{\partial H}{\partial p_i} \\[3mm] \dfrac{\partial p_i}{\partial t} = -\dfrac{\partial H}{\partial q_i} \end{cases} \tag{1.31}$$

One deduces

$$\begin{cases} q_i(t + dt) = q_i(t) + dt \cdot \dfrac{\partial H}{\partial p_i} \\[3mm] p_i(t + dt) = p_i(t) - dt \cdot \dfrac{\partial H}{\partial q_i} \end{cases} \tag{1.32}$$

and in particular

$$\begin{cases} \dfrac{\partial q_i(t + dt)}{\partial q_j(t)} = \delta_{ij} + dt \cdot \dfrac{\partial^2 H}{\partial q_j \partial p_i} \\[3mm] \dfrac{\partial p_i(t + dt)}{\partial p_j(t)} = \delta_{ij} - dt \cdot \dfrac{\partial^2 H}{\partial p_j \partial q_i} \end{cases} \tag{1.33}$$

The obtained determinant is developed to first order in dt :

$$\text{Det}(1 + dt \cdot \hat{M}) = 1 + dt \cdot \text{Tr } \hat{M} + O(dt^2)$$

with

$$\text{Tr } \hat{M} = -\sum_{i=1}^{3N} \frac{\partial^2 H}{\partial p_i \partial q_i} + \sum_{i=1}^{3N} \frac{\partial^2 H}{\partial p_i \partial q_i} = 0 \qquad (1.34)$$

Consequently, the volume around a point of the $6N$-dimensional phase space is conserved during the time evolution of this point; the number of systems in this volume, i.e., $D(\ldots q_i, \ldots, \ldots p_i, \ldots, t)d\tau$, is also conserved.

B. Number of systems versus time in a fixed volume $d\tau$ of the phase space

In the volume $d\tau = C_N \prod_{i=1}^{3N} dq_i \cdot dp_i$, at time t, the number of systems present is equal to

$$D(q_1, \ldots q_i, \ldots q_{3N}, p_1, \ldots p_i, \ldots p_{3N})d\tau \qquad (1.35)$$

Each system evolves over time, so that it may leave the considered volume while other systems may enter it.

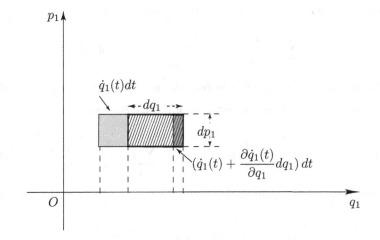

FIG. 1.6: In the volume $dq_1 \cdot dp_1$ (thick line), between the times t and $t + dt$, the systems which enter were in the light grey rectangle at t; those which leave the volume were in the dark grey hatched rectangle. It is assumed that $\dot{q}_1(t)$ is directed toward the right.

Take a "box", like the one in Fig. 1.6 which represents the situation for the only coordinates q_1, p_1. If only this couple of conjugated variables is considered, the problem involves a single coordinate q_1 in real space (so that the velocity $\dot{q}_1$ is necessarily normal to the "faces" at constant q_1) and we determine the variation of the number of systems on the segment for the q_1 coordinate $[q_1, q_1 + dq_1]$.

Considering this space coordinate, between the times t and $t + dt$, one calculates the difference between the number of systems which enter the volume through the faces normal to dq_1 and the number which leave through the same faces : the ones which enter were at time t at a distance of this face smaller or equal to $\dot{q}_1(t)dt$ (light grey area on the Fig.), their density is $D(q_1 - dq_1, \ldots q_i, \ldots q_{3N}, p_1, \ldots p_i, \ldots p_{3N})$. Those which leave were inside the volume, at a distance $\left(\dot{q}_1(t) + \dfrac{\partial \dot{q}_1(t)}{\partial q_1} dq_1 \right) dt$ from the face (dark grey hatched area).

The resulting change, between t and $t + dt$ and for this set of faces, of the number of particles inside the considered volume, is equal to

$$-\frac{\partial(\dot{q}_1(t)D)}{\partial q_1} dq_1 dt \tag{1.36}$$

For the couple of "faces" at constant p_1, an analogous evaluation is performed, also contributing to the variation of the number of particles in the considered "box." Now at $3N$ dimensions, for all the "faces," the net increase of the number of particles in $d\tau$ during the time interval dt is

$$-\sum_{i=1}^{3N} \frac{\partial(D\dot{q}_i(t))}{\partial q_i} dq_i dt - \sum_{i=1}^{3N} \frac{\partial(D\dot{p}_i(t))}{\partial p_i} dp_i dt \tag{1.37}$$

This is related to a change in the particle density in the considered volume through to $\dfrac{\partial D}{\partial t} dt d\tau$. Thus the net rate of change is given by

$$\frac{\partial D}{\partial t} = -\sum_{i=1}^{3N} \left(\frac{\partial D}{\partial q_i} \dot{q}_i + \frac{\partial D}{\partial p_i} \dot{p}_i \right) - \sum_{i=1}^{3N} D \left(\frac{\partial \dot{q}_i}{\partial q_i} + \frac{\partial \dot{p}_i}{\partial p_i} \right) \tag{1.38}$$

From the Hamilton equations, the factor of D is zero and the above equation is equivalent to

$$\frac{dD}{dt} = \frac{\partial D}{\partial t} + \sum_{i=1}^{3N} \left(\frac{\partial D}{\partial q_i} \dot{q}_i + \frac{\partial D}{\partial p_i} \dot{p}_i \right) = 0 \tag{1.39}$$

where $\dfrac{dD(q_1, \ldots, q_i, \ldots q_{3N}, p_1, \ldots, p_i, \ldots p_{3N})}{dt}$ is the time derivative of the probability density when one moves with the representative point of the system in the $6N$-dimensional phase space. This result is equivalent to the one of § A.

As an example, consider the microcanonical ensemble, in which the density D is a constant for the energies between E and $E + \delta E$, zero elsewhere. The density D is function of the sole energy, a constant in the microcanonical case, so that its derivative with respect to energy is zero between E and $E + \delta E$. Then D is independent of the coordinates q_i and of the momenta p_i, and from (1.39) it is time-independent.

Appendix 1.2

Properties of the Density Operator in Quantum Mechanics

The operator $\hat{D}$ defined by

$$\hat{D} = \sum_n p_n |\psi_n\rangle \langle \psi_n|$$

where p_n is the probability to realize the microstate $|\psi_n\rangle$, has the following properties :

i) it is hermitian : $\hat{D} = \hat{D}^+$

Indeed, for any operator $|u\rangle$,

$$\langle u|\hat{D}|u\rangle = \sum_n p_n |\langle u|\psi_n\rangle|^2 = \langle u|\hat{D}^+|u\rangle \tag{1.40}$$

ii) it is defined positive

$$\langle u|\hat{D}|u\rangle = \sum_n p_n |\langle u|\psi_n\rangle|^2 \geq 0 \tag{1.41}$$

iii) it is normed to unity, since

$$\text{Tr}\,\hat{D} = \sum_n p_n = 1 \tag{1.42}$$

Appendix 1.3

Estimation of the Number of Microstates for Free Particles Confined Within a Volume Ω

Take the example of a *free point particle*, of energy and momentum linked by $\dfrac{p^2}{2m} = \varepsilon$; in three dimensions, the particle occupies the volume Ω. The constant energy surfaces correspond to a constant p. In the 6-dimensional $(\vec{r}, \vec{p})$ phase space for a single particle, the volume associated with the energies between ε and $\varepsilon + \delta\varepsilon$ is equal to : $\Omega 4\pi p^2 \delta p = \Omega 4\pi p.p\delta p$, with

$$\delta\varepsilon = \frac{p\delta p}{m}, \text{ that is, } \delta p = \sqrt{\frac{m}{2\varepsilon}}\delta\varepsilon \qquad (1.43)$$

For a single particle, the number of accessible states between ε and $\varepsilon + \delta\varepsilon$ is proportional to $\Omega p \cdot p\delta p$, i.e., $\Omega \varepsilon^{1/2}\delta\varepsilon$, i.e., also to $\Omega \varepsilon^{(3/2)-1}\delta\varepsilon$.

For N independent *free particles*, the constant-energy sphere of the $6N$-dimensional phase space contains all the states of energy smaller or equal to E, where E is of the order of $N\varepsilon$. The volume of such a sphere varies in $\Omega^N p^{3N}$. The searched microstates, of energy between E and $E + \delta E$, correspond to a volume of the order of $\Omega^N p^{3N-1}\delta p$, that is, $\Omega^N p^{3N-2}\delta E$, for $p\delta p \propto \delta E$. The number $W(E)$ of microstates with energy between E and $E + \delta E$ is proportional to this volume, i.e.,

$$W(E) = \left(\frac{\Omega}{\Omega_0}\right)^N \left(\frac{E}{E_0}\right)^{3N/2} \frac{\delta E}{E} \qquad (1.44)$$

where the constants Ω_0 and E_0 provide the homogeneity of the expression and where δE, the uncertainty on the energy E, is very much smaller than this energy value.

The increase of $W(E)$ versus energy and volume is *extremely fast* : for example, if N is of the order of the Avogadro number, $\mathcal{N}$,

$$W(E) = \left(\frac{\Omega}{\Omega_0}\right)^{6\times10^{23}} \left(\frac{E}{E_0}\right)^{9\times10^{23}} \frac{\delta E}{E} \tag{1.45}$$

which is such a huge number that it is difficult to imagine it! The logarithm in the definition of the Boltzmann entropy S is here :

$$\ln W(E) = N \ln \frac{\Omega}{\Omega_0} + \left(\frac{3N}{2}\right) \ln \frac{E}{E_0} + \ln \frac{\delta E}{E} \tag{1.46}$$

E is of the order of N times the energy of an individual particle. For N in the $\mathcal{N}$ range, the dominant terms in $\ln W(E)$ are proportional to N : indeed for an experimental accuracy $\delta E/E$ of the order of 10^{-4}, the last logarithm is equal to -9.2 and is totally negligible with respect to the two first terms, of the order of 10^{23}.

Note : the same type of calculation, which estimates the number of accessible states in a given energy range, will also appear when calculating the density of states for a free particle ($\S\,6.4$).

Chapter 2

The Different Statistical Ensembles. General Methods in Statistical Physics

In the first chapter we convinced ourselves that the macroscopic properties of a physical system can only be analyzed through a statistical approach, in the framework of Statistical Physics.

Practically, the statistical description of a system with a very large number of particles in equilibrium, and the treatment of any *exercise* or *problem* of Statistical Physics, *always* follow the two steps (see the section General method for solving exercises and problems, at the end of Ch. 9) :

– *First, determination of the microscopic states* (microstates) accessible by the N-particles system : this is a Quantum Mechanics problem, which may be reduced to a Classical Mechanics one in some specific cases. § 2.1 will schematically present the most frequent situations.

– *Second, evaluation of the probability for the system to be in a particular microstate*, in the specified physical conditions : it is at this point that the statistical description comes in.

To solve this type of problem, one has to express that *in equilibrium the statistical entropy is maximum, under the constraints* defined by the physical situation under study : the system may be isolated, or in thermal contact

31

with a heat reservoir (a very large system, which will dictate its temperature), the system may exchange particles or volume with another system,
and so forth. The experimental conditions define the constraint(s) from the
conservation laws adapted to the physical situation (conserved total energy,
conserved number of particles, etc.). In this chapter, the most usual statistical
ensembles will be presented : "microcanonical" ensemble in § 2.2, "canonical"
ensemble in § 2.4, "grand canonical" ensemble in § 2.5, using the names introduced at the end of the 19th century. In all these statistical ensembles,
one reproduces the considered system in a thought experiment, thus realizing
an assembly of macroscopic objects built under the same initial conditions,
which allows average values of fluctuating physical parameters to be defined.
For a given physical condition, one will decide to use the statistical ensemble
most adapted to the analysis of the problem.

However, as will be shown, for systems with a *macroscopic* number of particles
(the only situation considered in this book), there is equivalence, to a extremely small relative fluctuation, between the various statistical ensembles that
will be described here : in the case of a macroscopic system, a measurement will
not distinguish between a situation in which the energy is fixed and another
one in which the temperature is given. It is only the convenience of treatment
of the problem that will lead us to choose one statistical ensemble rather than
another. (On the contrary for systems with a small number of particles, or
of size intermediate between the microscopic and macroscopic ranges, the so-
called "mesoscopic systems," nowadays much studied in Condensed Matter
Physics, the equivalence between statistical ensembles is not always valid.)

2.1   Determination of the Energy States of an N-Particle System

For an individual particle i in translation, of momentum $\vec{p}_i$ and mass m,
under a potential energy $V_i(\vec{r}_i)$, the time-independent Schroedinger equation
(eigenvalues equation) is

$$\hat{h}_i|\psi_i^{\alpha_i}\rangle = \left(\frac{\hat{\vec{p}}_i^{\,2}}{2m} + V_i(\vec{r}_i)\right)|\psi_i^{\alpha_i}\rangle$$
$$= \varepsilon_i^{\alpha_i}|\psi_i^{\alpha_i}\rangle \tag{2.1}$$

where α_i expresses the different states accessible to this particle.

As soon as several particles are present, one has to consider their interactions,
which leads to a several-body potential energy. Thus, the hamiltonian for N
particles in motion, under a potential energy and mutual interactions, is given

by

$$\hat{H} = \sum_i \frac{\hat{p_i}^2}{2m} + V_i(\vec{r}_i) + \sum_{i<j} V_{ij}(\vec{r}_i, \vec{r}_j) \qquad (2.2)$$

This is the case, for example, for the molecules of a gas, confined within a limited volume and mutually interacting through Van der Waals interactions; this is also the situation of electrons in solids in attractive Coulombic interaction with the positively charged nuclei (V_i term), and in repulsion between themselves (V_{ij} term). The latter interactions, expressed in the V_{ij}'s, are essential in the approach toward thermal equilibrium through the transitions they induce between microscopic states, but they complicate the solution of the N-particle eigenstate problem. Now they mostly correspond to energy terms very small with respect to those associated with the remaining part of the hamiltonian.

An analogous problem is faced when one is concerned by the magnetic moments of a ferromagnetic material : the hamiltonian of an individual electronic intrinsic magnetic moment located in a magnetic field is given by

$$\hat{h} = -\hat{\vec{\mu}}_B \cdot \vec{B} \qquad (2.3)$$

where the Bohr magneton operator $\hat{\vec{\mu}}_B$ is related to the spin operator $\hat{\vec{S}}$ by :

$$\hat{\vec{\mu}}_B = -\frac{e}{m}\hat{\vec{S}} \qquad (2.4)$$

and has for eigenvalues $\mp e\hbar/2m$ (see a course of Quantum Mechanics). The hamiltonian of an ensemble of magnetic moments in a magnetic field $\vec{B}$ contains, in addition to terms similar to $\hat{h}$, other terms expressing their mutual interactions, i.e.

$$\hat{H} = -\sum_i \hat{\vec{\mu}}_{Bi} \cdot \vec{B} + \sum_{i<j} J_{ij} \hat{\vec{\mu}}_{Bi} \cdot \hat{\vec{\mu}}_{Bj} \qquad (2.5)$$

where J_{ij} is the coupling term between the moments localized at sites i and j. According to the site i or j, the moment orientation is different.

All these types of hamiltonians, in which several-particle terms appear, are generally treated through approximations, which allow them to be reduced to the simplest situation where $\hat{H}$ is written as a *sum of similar hamiltonians*, each concerning a single particle. There are several methods to achieve such a reduction :

– either neglecting the interactions between particles in the equilibrium state, as they correspond to very small energies. As a consequence, in the above example of the gas molecules, the only term included in the potential energy expresses the confinement with a box (confinement potential) or an external field (gravity); the particles are then considered as "free";

– or using the so-called "mean field" treatment : each particle is subjected to a mean effect from the other ones. This will be a repulsive potential energy in the case of the Coulombic interaction between electrons in a solid, an effective magnetic field equivalent to the interactions from the other magnetic moments in the case of a ferromagnetic material;

– or changing of variable in the hamiltonian $\hat{H}$, in order to decouple the variables, so that $\hat{H}$ may be re-written as a sum of terms ("normal modes" defining quasi-particles, for example, in the problem of the atomic vibrations in a crystal at a given temperature).

From now on, let us assume that the total-system hamiltonian is indeed expressed as the sum of similar individual-particle hamiltonians

$$\hat{H} = \sum_i \hat{h}_i \ , \quad \text{with } \hat{h}_i \psi_i^{\alpha_i}(\vec{r}_i) = \varepsilon_i^{\alpha_i} \psi_i^{\alpha_i}(\vec{r}_i) \tag{2.6}$$

Then it will only be necessary to solve the problem for an individual particle : indeed you learnt in Quantum Mechanics that the solution of

$$\hat{H}\psi(\vec{r}_1, \vec{r}_2, \dots, \vec{r}_N) = E\psi(\vec{r}_1, \vec{r}_2, \dots, \vec{r}_N) \tag{2.7}$$

is

$$\Big(\sum_j \hat{h}_j\Big)\Big(\prod_{i=1}^{N} \psi_i^{\alpha_i}(\vec{r}_i)\Big) = \Big(\sum_j \varepsilon_j^{\alpha_j}\Big)\Big(\prod_{i=1}^{N} \psi_i^{\alpha_i}(\vec{r}_i)\Big) \tag{2.8}$$

i.e.,

$$E = \sum_i \varepsilon_i^{\alpha_i} \ , \ \psi(\vec{r}_1, \vec{r}_2, \dots, \vec{r}_N) = \prod_{i=1}^{N} \psi_i^{\alpha_i}(\vec{r}_i) \tag{2.9}$$

The total energy of the system is the sum of the individual particle energies and the N-particle eigenfunction is the product of the eigenfunctions for each single particle.

In the first three chapters of this book, the problems studied will concern discrete variables (for example,, spin magnetic moments) or particles distinguishable by their *fixed* coordinates, like the atoms of a solid vibrating around their equilibrium position at finite temperature. The systems with translation degrees of freedom will be analyzed in the classical framework in chapter 4 (ideal gas), or in the framework of the Quantum Statistics from chapter 5 until the end of the course : in the Quantum Statistics description of the accessible microstates, one has to account for the indistinguishability of the particles, as expressed by the Pauli principle.

In all that follows, we will assume that the microscopic states of the considered system are known. We are going to rely on the hypothesis of maximum entropy, as introduced in chapter 1, § 1.3 and § 1.4, under the specific physical constraints due to the given experimental conditions ; we will then deduce the system macroscopic properties, using Statistical Physics.

2.2 Isolated System in Equilibrium : "Microcanonical Ensemble"

In such a system one assumes that there is no possible exchange with its surroundings (Fig. 2.1) : the volume is fixed, the number of particles N is given, the total energy assumes a fixed value E, i.e., its lies in the range between E $E + \delta E$, where δE is the uncertainty. This situation is called "*microcanonical ensemble.*" It is the framework of application of the Boltzmann hypothesis of equiprobability of the $W(E)$ accessible microstates (§ 1.3.3). The probability to realize any of these microstates is equal to $1/W(E)$; the statistical entropy S is maximum under these conditions (with respect to the situation where the probabilities would not be equal) and is given by

$$S = k_B \ln W(E) \tag{2.10}$$

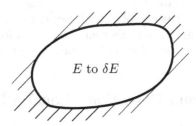

FIG. 2.1 : Isolated system. Its energy has a fixed value.

To solve this type of problem, one has to determine $W(E)$: this means performing a combinatory calculation of the number of occurrences of the macroscopic energy E from the various configurations of the microscopic components of the system. For example, for an ensemble of localized magnetic moments located in an external magnetic field, to each fixed value of E corresponds a value of the macroscopic magnetization. The corresponding number of microstates $W(E)$ was evaluated in § 1.3.1. The other physical macroscopic parameters are deduced from S and its partial derivatives (see § 3.5). In particular a microcanonical temperature is defined from the partial derivative of S versus $E : 1/T = \partial S/\partial E$.

2.3   Equilibrium Conditions for Two Systems in Contact, Only Exchanging Energy

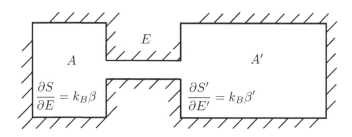

FIG. 2.2: The systems A and A' only exchange energy, the combined system $A_0{=}A{+}A'$ is isolated.

Two distinct systems A and A' are brought into thermal contact. The system A_0, made up of the combination of A and A' (Fig. 2.2) is the isolated system, to which we can apply the above approach, i.e. look for the maximum of its statistical entropy. The systems in contact, A and A', can only exchange energy and are weakly coupled, which means that a possible interaction energy is neglected. We can also say that the system A under study has the constraint to be coupled to A', the total energy $E + E'$ being conserved. This is a very common situation, like the one of a drink can in a refrigerator, the ensemble (can-refrigerator) being taken as thermally insulated from the room in which the refrigerator stands, the room representing the surrondings.

2.3.1   Equilibrium Condition : Equal β Parameters

The total energy and the total statistical entropy are shared between A and A' :

$$E_0 = E + E' \tag{2.11}$$

$$S_{A_0}(E_0, E) = S_A(E) + S_{A'}(E_0 - E) \tag{2.12}$$

From the Boltzmann relation (2.10), the equation (2.12) is equivalent to

$$W_0(E_0, E) = W_A(E) \cdot W_{A'}(E_0 - E) \tag{2.13}$$

since for weakly coupled systems the number of microstates of the combined system is the product of the numbers of microstates of the two subsystems.

After a long enough time has elapsed, the combined system (can-refrigerator) A_0 reaches an equilibrium situation. Since the combination A_0 of the two systems A and A' is an isolated system, in equilibrium its entropy is maximum with respect to the exchange of energy between A and A' :

$$S_{A_0}(E_0, E) = S_A(E) + S_{A'}(E_0 - E) \tag{2.14}$$

$$\frac{\partial S_{A_0}(E_0, E)}{\partial E} = 0 , \ \text{i.e.,} \ \frac{\partial S_{A_0}(E_0, E)}{\partial E} = \frac{\partial S_A(E)}{\partial E} + \frac{\partial S_{A'}(E_0 - E)}{\partial E} = 0 \tag{2.15}$$

Once the equilibrium has been reached :

$$\frac{\partial S_A(E)}{\partial E} = -\frac{\partial S_{A'}(E_0 - E)}{\partial E} = \frac{\partial S_{A'}(E')}{\partial E'} \tag{2.16}$$

The obtained equilibrium condition is that the *partial derivative of entropy versus energy* takes the *same value* in *both systems* in contact, A and A', all the other parameters (volume, etc.) being kept constant. This derivative will be written :

$$\frac{\partial S}{\partial E} = k_B \beta \tag{2.17}$$

where k_B is the Boltzmann constant.

In equilibrium, reached for an energy value of system A equal to $E = \tilde{E}$, one thus has

$$\beta = \beta' \tag{2.18}$$

We know that the entropy S and the energy E are both *extensive** parameters : by definition the values of extensive parameters are doubled when the system volume Ω is doubled together with the number N of particles, while the density N/Ω, an *intensive** parameter, is maintained in such a transformation. In such a process, the parameter β is not changed, it is an *intensive* parameter having the dimension of reciprocal energy, the dimension of $k_B \beta$ being a reciprocal temperature. In § 3.3 the partial derivative of the entropy versus energy will be identified with $1/T$, where T is the absolute temperature, so that (2.18) expresses that in equilibrium $T = T'$.

2.3.2 Fluctuations of the Energy Around its Most Likely Value

The number of microstates of the total system A_0 realizing the splitting of the energies [E in A, $E_0 - E$ in A'] is proportional to the product

$W_A(E) \cdot W_{A'}(E_0 - E)$. The probability of occurrence of the corresponding microstate is

$$p(E, E_0 - E) = \frac{W_A(E)W_{A'}(E_0 - E)}{\sum_E W_A(E)W_{A'}(E_0 - E)} = \frac{W_A(E)W_{A'}(E_0 - E)}{W_{\text{tot}}(E_0)} \qquad (2.19)$$

Let us consider the *variation* of this probability versus E *around the value* $\tilde{E}$ achieving the entropy maximum with respect to an energy exchange between A and A'. In (2.19), only the numerator is a function of E, since the denominator is a sum over all the possible values of E.

For a macroscopic system, $W(E)$ increases very fast with E, as already verified on the example of free particles in Appendix 3 of Chapter 1. On the other hand, $W_{A'}(E_0 - E)$ strongly decreases with E if A' is macroscopic. Their product goes through a very steep maximum, corresponding to the state of maximum probability of the combined system, reached for $E = \tilde{E}$. This energy is very close to the average value $\langle E \rangle$ of the energy in system A, as the probability practically vanishes outside the immediate vicinity of $\tilde{E}$. The probability variation around $\tilde{E}$ is much faster than that of its logarithm, which varies versus E in the same way as

$$S_{A_0}(E) = k_B \ln[W_A(E)W_{A'}(E_0 - E)] \qquad (2.20)$$

because the combined system A_0 is isolated.

Note that in the particular case of a free-particle system, where $W(E) \sim E^N$ (see Appendix 3, Chapter 1), $\beta = \left(\dfrac{\partial S}{\partial E}\right)_\Omega$ is of the order of N/E. Since the total energy is of the order $\tilde{E}$, the average energy per particle is of the order $1/\beta = k_B T$, to a numerical factor of order unity. The average energy of a free particle at temperature T is thus of order $k_B T$.

Now we look for the probability *variation* versus the energy of system A for E close to the most likely value $\tilde{E}$. The logarithm of this probability is proportional to the entropy of the combined system $S_{A_0}(E)$, considered as a function of E, as A_0 is an isolated system to which the Boltzmann formula $S_{A_0}(E) = k_B \ln W_0(E)$ applies.

In equilibrium, the entropy of the combined system A_0 is maximum with respect to an energy exchange between the two systems; the entropy of A_0 can be developed around this maximum, for $E - \tilde{E}$ small :

$$S_{A_0}(E) = S_{A_0}(\tilde{E}) + \left(\frac{\partial S_{A_0}}{\partial E}\right)_{\tilde{E}}(E - \tilde{E}) + \frac{1}{2}\left(\frac{\partial^2 S_{A_0}}{\partial E^2}\right)_{\tilde{E}}(E - \tilde{E})^2 \qquad (2.21)$$
$$+ O\big((E - \tilde{E})^3\big)$$

Here the first derivative is equal to zero and the second derivative is negative, as the extremum is a maximum. This second derivative includes the contributions of both A and A' and is equal to :

$$\frac{\partial^2 S_{A_0}}{\partial E^2} = \frac{\partial^2 S_A}{\partial E^2} + \frac{\partial^2 S_{A'}}{\partial E'^2} = k_B \left(\frac{\partial \beta}{\partial E} + \frac{\partial \beta'}{\partial E'} \right) = -k_B \lambda_0 \qquad (2.22)$$

All these terms are calculated for $E = \tilde{E}$. In fact, since $\beta = 1/k_B T$, stating that $\partial \beta / \partial E$ is negative means that the internal energy increases with temperature : this is always realized in practical cases.[1].

From the entropy properties (2.21) and (2.22), using the Boltzmann relation one deduces those of the number of accessible microstates :

$$W_0(E) \approx W_0(\tilde{E}) \exp \left[-\lambda_0 (E - \tilde{E})^2/2 \right] \qquad (2.23)$$

The probability of realizing the state with the sharing of the energies [E in system A and $E_0 - E$ in system A'] is proportional to $W_0(E)$. It is given by

$$p(E, E_0 - E) \approx p(\tilde{E}, E_0 - \tilde{E}) \exp \left(-\lambda_0 (E - \tilde{E})^2/2 \right) \qquad (2.24)$$

This is a Gaussian function around the equilibrium state $\tilde{E}$. Its width $\Delta E = 1/\sqrt{\lambda_0}$ depends of the size of the total system like

$$\left(\frac{\partial \beta}{\partial E} + \frac{\partial \beta'}{\partial E'} \right)^{-1/2} \qquad (2.25)$$

that is, like the square root of a number of particles, since E and E' are extensive and the parameters β and β' intensive. The *relative* width $\Delta E / \tilde{E}$ varies as a reciprocal number of particles. For a macroscopic system, where N and $N' \sim 10^{23}$, this relative width is of the order of 10^{-11}, thus unmeasurable : the fluctuations of the energy of system A around its equilibrium value are practically null. This is equivalent to saying that, for a system with a very large number of particles, situations $i)$, in which the energy of A is exactly equal to the value $\tilde{E}$, and $ii)$, in which system A reached an equilibrium energy $\tilde{E}$ through exchanges with a heat reservoir, lead to the same physical measurement of the energy.

---

[1] When the total energy has an upper bound, as in the case of the spins in a paramagnetic solid, the maximum energy is reached when all the magnetic moments are antiparallel to the applied magnetic field. In the vicinity of this maximum, adding energy produces a decrease in entropy, so that (2.17) leads to a negative β, thus to a " negative temperature." However in the situation of a *real* paramagnetic solid, the spin degree of freedom is in equilibrium with the other degrees of freedom, like the ones related to the atoms' vibrations around their equilibrium positions. For the latter ones, the energy is not limited and for the whole solid, the total energy indeed increases with temperature.

2.4　System in Contact with a Heat Reservoir, "Canonical Ensemble"

When two systems are brought into thermal contact, one being much larger than the other (the can is much smaller than the refrigerator) (Fig. 2.3), the larger system behaves like a heat reservoir of energy $E_0 - E$ close to E_0, the energy E of the other system thus being very small. The statistical entropy, and, consequently, the parameter β of the larger system, are close to those of the total system :

$$\left(\frac{\partial S_{A'}}{\partial E'}\right)_{E_0 - E} = \left(\frac{\partial S_{A'}(E)}{\partial E}\right)_{E_0} - E\left(\frac{\partial^2 S_{A'}(E)}{\partial E^2}\right)_{E_0} + O(E^2) \qquad (2.26)$$

$$= k_B \beta_0 - E\left(\frac{\partial^2 S_{A'}(E)}{\partial E^2}\right)_{E_0} + O(E^2) \qquad (2.27)$$

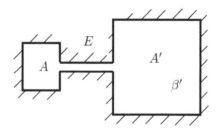

FIG. 2.3: The system A is in contact with the heat reservoir A' which dictates its parameter β' to A.

When the equilibrium is reached, the parameter β of the smaller system is adjusted to the parameter β' of the larger system, which itself takes a value close to that of the parameter β_0 of the ensemble : the larger system acts as a *heat reservoir* or *thermostat*, a body that keeps its value of β and dictates this parameter to the smaller system (we will see in chapter 3 that this is equivalent to saying that it keeps its temperature when put in thermal contact : for example, in equilibrium the can takes the temperature of the refrigerator). The smaller system A can be a macroscopic system, sufficiently smaller than A', or even a single microscopic particle, like an individual localized magnetic moment inside a paramagnetic solid, with respect to which the rest of the macroscopic solid plays the role of a reservoir.

The situation in which the system under study is in thermal contact with a heat reservoir that gives its temperature T to the system is called "canonical," since it is a very frequent situation among the studied problems. The ensemble of systems in contact with the heat reservoir, on which we will calculate the ensemble average, is the "canonical ensemble."

2.4.1 The Boltzmann Factor

We are thus in the situation where a smaller system A is coupled to a larger system A' that dictates its value of the parameter β, which is close to the value for the combined isolated system.

First consider the probability p_i that *a specific microstate i* of system A, of energy E_i, is produced. It only depends on the properties of the *heat reservoir* and is the ratio of the number of microstates $W_{\text{res}}(E_0 - E_i)$ of the reservoir with this energy to the sum of all the numbers of microstates for all the reservoir energies $E_0 - E_i$:

$$p_i = \frac{W_{\text{res}}(E_0 - E_i)}{\sum_i W_{\text{res}}(E_0 - E_i)} \tag{2.28}$$

$W_{\text{res}}(E_0 - E_i)$ is calculated in the microcanonical ensemble for the given energy $E_0 - E_i$ and is related to $S_{\text{res}}(E_0 - E_i)$ through the Boltzmann relation

$$\ln W_{\text{res}}(E_0 - E_i) = \frac{1}{k_B} S_{\text{res}}(E_0 - E_i) \tag{2.29}$$

Indeed, once the reservoir energy $E_0 - E_i$ is fixed, the different microstates in this situation are equally likely.

Since

$$S_{\text{res}}(E_0 - E_i) = S_{A'}(E_0) - E_i \left(\frac{\partial S_{A'}}{\partial E'} \right)_{E'=E_0} + \ldots \tag{2.30}$$

$$\ln W_{\text{res}}(E_0 - E_i) = \ln W_{A'}(E_0) - E_i \beta_0 + \ldots \tag{2.31}$$

$$= \text{constant} - \beta_0 E_i + \ldots$$

the probability of occurrence of this microstate is proportional to $W_{\text{res}}(E_0 - E_i)$, i.e.,

$$p_i = C \exp(-\beta_0 E_i) \tag{2.32}$$

In the probability p_i there appears the exponential of the energy E_i of the system. This is the so-called "Boltzmann factor" (1871), or "canonical distribution," an expression that will be used very often in this course, every time we will study an *ensemble of distinguishable particles*, in *fixed number N*, of *given temperature* $T = \dfrac{1}{k_B \beta}$ (the case of indistinguishable particles will

be treated in the framework of the Quantum Statistics in chapter 5 and the
following ones). As an example, Fig. 2.4 schematizes the Boltzmann factor for
a 4-microstate system.

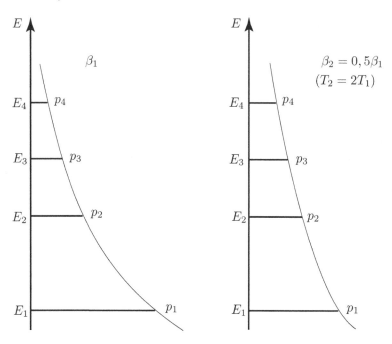

FIG. 2.4: The considered system has four microstates of energies E_1 to E_4.
The abscissae are proportional to the probabilities of occurrence of one of the
microstates, for two different values of β (or of the temperature).

If the *total number* of microstates of system A corresponding to the *same
energy E* is $W(E)$, then the probability that system A, coupled to A', *has the
energy E*, is equal to :

$$p(E) = C'W(E)\exp(-\beta_0 E) \tag{2.33}$$

where $W(E)$ is the degeneracy (the number of ways to realize this energy)
of the state of energy E. The probability $p(E)$ is the product of the function
$W(E)$, very rapidly increasing with energy for a macroscopic system, by a
decreasing exponential : the probability has a very steep maximum for $E = \tilde{E}$.

2.4.2   Energy, with Fixed Average Value

We now show that another condition, i.e., the constraint of a given average
value $\langle E \rangle$ of the energy of a system A, leads to a probability law analogous to

the one just found. The only difference is that the value of β, now a parameter, should be adjusted so that the average of energy indeed coincides with the given value $\langle E \rangle$.

We have to maximize the statistical entropy of A while the value $\langle E \rangle$ is given. We thus have to realize

$$S_A = -k_B \sum_i p_i \ln p_i \text{ maximum, under the constraints}$$

$$\begin{cases} \sum_i p_i = 1 \\ \sum_i p_i E_i = \langle E \rangle \end{cases} \tag{2.34}$$

A general mathematical method allows one to solve this type of problem, it is the method of Lagrange multipliers explained in Appendix 2.1 : parameters are introduced, which are the Lagrange multipliers (here β). These parameters are adjusted at the end of the calculation, in order that the average values be equal to those given by the physical properties of the system, here the value of the average energy.

One thus obtains for the probability of occurrence of the microstate of energy E_i :

$$p_i = \frac{\exp(-\beta E_i)}{\sum_i \exp(-\beta E_i)} \tag{2.35}$$

that is, an expression of the "Boltzmann-factor" type (2.32) but here β does not characterize the temperature of a real heat reservoir. It is rather determined by the condition that the average energy of the system A should be the one given by the problem conditions, that is :

$$\frac{\sum_i E_i \exp(-\beta E_i)}{\sum_i \exp(-\beta E_i)} = \langle E \rangle \tag{2.36}$$

This method of the Lagrange multipliers is followed in some Statistical Physics courses to obtain the general expression of the Boltzmann factor (for example, those given at Ecole polytechnique by R. Balian, E. Brézin, or A. Georges and M. Mézard).

2.4.3   Partition Function Z

The probability for the system in contact with a heat reservoir to be in a state of energy E_i is

$$p_i = \frac{e^{-\beta E_i}}{Z_N} \, , \ \ \text{with} \beta = \frac{1}{k_B T} \tag{2.37}$$

The term

$$Z_N = \sum_i e^{-\beta E_i} \tag{2.38}$$

which allows one to norm the Boltzmann factors is called the "canonical partition function for N particles" (in German "Zustandssumme," which means "sum over the states").

Formulae (2.37) and (2.38) are among the most important ones of this course!

The complete statistical information on the considered problem is contained in its partition function, since from Z and its partial derivatives all the average values of the physical macroscopic parameters of the system can be calculated. We will show this property now on the examples of the average energy and the energy fluctuation around its average (for the entropy see § 2.4.5). Indeed, the average energy is given by

$$\begin{aligned} \langle E \rangle &= \sum_i p_i E_i = \sum_i E_i \frac{\exp(-\beta E_i)}{Z} \\ &= -\frac{1}{Z}\frac{\partial Z}{\partial \beta} = -\frac{\partial \ln Z}{\partial \beta} \end{aligned} \tag{2.39}$$

It appears here that $\ln Z$ is extensive like $\langle E \rangle$.

To determine the energy fluctuation around its average value, one calculates

$$(\Delta E)^2 = \langle E - \langle E \rangle \rangle^2 = \langle E^2 \rangle - \langle E \rangle^2 \tag{2.40}$$

A procedure similar to that of (2.39) is followed to find $\langle E^2 \rangle$:

$$\langle E^2 \rangle = \sum_i p_i E_i^2 = \sum_i E_i^2 \frac{\exp(-\beta E_i)}{Z} = \frac{1}{Z}\frac{\partial^2 Z}{\partial \beta^2} \tag{2.41}$$

$$(\Delta E)^2 = \frac{1}{Z}\frac{\partial^2 Z}{\partial \beta^2} - \frac{1}{Z^2}\left(\frac{\partial Z}{\partial \beta}\right)^2 = \frac{\partial}{\partial \beta}\left(\frac{1}{Z}\frac{\partial Z}{\partial \beta}\right) = \frac{\partial^2 \ln Z}{\partial \beta^2} \tag{2.42}$$

One verifies here that, since $\ln Z$ varies as the number of particles of the system and β is intensive, ΔE varies as $N^{1/2}$ and $\Delta E/E$ as $N^{-1/2}$ (in agreement with the result of § 2.3.2), which corresponds to an extremely small value in a macroscopic system : for $N = 10^{23}$, $\Delta E/E$ is of the order of 10^{-11} !

Note 1 : A demonstration (outside the framework of the present course, see, for example, R. Balian chapter 5 § 5) can be done of the equivalence between all the statistical ensembles in the case of macroscopic systems : it is valid in the "thermodynamical limit," which consists in simultaneously having several parameters tending toward infinity : the volume Ω (which means in particular that it is very much larger than typical atomic volumes), the particle number N and the other extensive parameters, while keeping constant the particle density N/Ω and the other intensive parameters. In such a limit, it is equivalent to impose an exact value of a physical parameter (like the energy in the microcanonical ensemble) or an average value of the same parameter (the energy in the canonical ensemble).

In fact, the technique is simpler in the canonical ensemble, where $\ln Z$ is calculated on *all the states* without any restriction, than in the microcanonical ensemble, where the calculation of $\ln W(E)$ requires the limitation to the range of energies between E and $E + \delta E$. Besides, for macroscopic systems one understands that the predicted physical results, using either the microcanonical or the canonical statistical ensemble, cannot be distinguished through measurements.

Note 2 : The value of Z_N "gives a hint" at the number of microstates that can be achieved at the experiment temperature. Indeed at very low temperature, where $\beta E_i \to \infty$, only the fundamental state E_0 is realized and $Z_N = 1$. On the other hand, at high temperature where $\beta E_i \to 0$, many terms are of the order of unity, corresponding to practically equally likely states.

2.4.4 Entropy in the Canonical Ensemble

We just showed that, using the partition function, we can calculate the average energy of system A at temperature T, together the fluctuation around this average energy. We now have to express the statistical entropy of A in these conditions. Here the probabilities of occurrence of the different microstates are not equal, since they depend of their respective energies through Boltzmann factors. Thus the definition of the statistical entropy to be used is no longer that of Boltzmann, but rather the one of Gibbs :

$$S_A = -k_B \sum_i p_i \ln p_i \tag{2.43}$$

with

$$p_i = \frac{1}{Z} e^{-\beta E_i} \;,\; \sum_i p_i = 1 \tag{2.44}$$

Consequently,

$$S_A = -k_B \sum_i p_i (-\ln Z - \beta E_i) \tag{2.45}$$

$$S_A = k_B (\ln Z + \beta \langle E \rangle) = k_B \ln Z + k_B \beta U \tag{2.46}$$

The average energy $\langle E \rangle$ of the macroscopic system is identified to the internal energy U of Thermodynamics (see § 3.2).

Note that, for a large system in which the energy fluctuation is relatively very small around $\langle E \rangle$, the probability of occurrence of this energy value is very close to unity. The number of occurrences of this average energy is $W(\langle E \rangle)$, so that the Gibbs entropy (2.43) practically reduces to the Boltzmann entropy (2.10) for this value $\langle E \rangle$.

2.4.5   Partition Function of a Set of Two Independent Systems with the Same β Parameter (Same T)

In § 2.1 we saw that, in the N-particle problems solved in this book, we always decompose the hamiltonian of the total system into a sum of hamiltonians for individual particles; a particle may have several degrees of freedom, whence another sum of hamiltonians. The energy of a microstate is thus a sum of energies and now we see the consequence of this property on the partition function of such a system in thermal equilibrium.

This situation is schematized by simply considering a system A made up of two independent subsystems A_1 and A_2, both in contact with a heat reservoir at temperature T. A particular microstate of A_1 has the energy E_{1i}, while the energy of A_2 is E_{2j} and the energy of A is E_{ij}. Then

$$\begin{cases} E_{ij} = E_{1i} + E_{2j} \\[2mm] Z = Z_1 Z_2 = \sum_{ij} \exp\{-\beta(E_{1i} + E_{2j})\} \;,\; \text{with } \beta = \frac{1}{k_B T} \\[2mm] \ln Z = \ln Z_1 + \ln Z_2 \\[2mm] \langle E \rangle = \langle E_1 \rangle + \langle E_2 \rangle \\[2mm] S = S_1 + S_2 \end{cases} \tag{2.47}$$

This property is very often used : when the *energies* of two independent systems at the same temperature *sum*, the corresponding *partition functions multiply*. Consequently, in the case of distinguishable independent particles with several degrees of freedom, one will separately calculate the partition functions for the various degrees of freedom of an individual particle; then one will multiply the factors corresponding to the different degrees of freedom and particles to obtain the partition function Z of the total system. Finally, in the special case of the ideal gas, one will introduce the factor C_N into Z (see § 4.4.3 and 6.7).

Take the example of the vibrations of atoms around their equilibrium positions, owing to thermal motion, in a solid in thermal equilibrium at temperature T : this is a very classical exercise and here we only sketch its solution. In the model proposed by Einstein (1907), one assumes that the atoms, in number N, are points and that they are all attracted toward their respective equilibrium position with the same restoring constant, that is, with the same frequency ω. The value of ω depends of the mechanical properties of the solid, in particular of its stiffness. Then the hamiltonian for a particular atom in motion is given by

$$\hat{h}_i = \frac{\hat{\vec{p}}_i^2}{2m} + \frac{1}{2}m\omega^2\hat{\vec{r}}_i^2 = \hat{h}_{xi} + \hat{h}_{yi} + \hat{h}_{zi} \tag{2.48}$$

It is the sum of three hamiltonians of the "harmonic-oscillator" type, identical for each coordinate. The eigenvalues of the hamiltonian $\hat{h}_{xi}$ relative to the coordinate x of site i are

$$E_{xi}^n = \left(n_{xi} + \frac{1}{2}\right)\hbar\omega \tag{2.49}$$

The contribution to the partition function of this degree of freedom is equal to

$$z_{xi} = \sum_n \exp(-\beta(n_{xi} + 1/2)\hbar\omega) \tag{2.50}$$

The partition function for the total system is the product of $3N$ terms similar to this one. The average energy is deduced using (2.39), it is equal to $3N$ times the average value for the coordinate x of site i.

The parameter accessible to experiment is the specific heat at constant volume C_v, defined as the derivative of the average energy with respect to temperature. The above solution gives for this lattice contribution

$$C_v = \frac{3Nk_B(\Theta/T)^2}{(e^{\Theta/T} - 1)^2} \ , \ \text{taking} \ \frac{\hbar\omega}{k_B} = \Theta \tag{2.51}$$

This expression is sketched on Fig. 2.5.

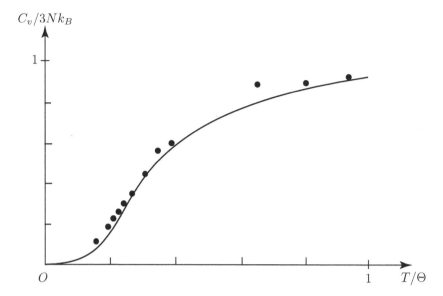

FIG. 2.5: Variation of C_v versus temperature, according to the Einstein model. The dots are experimental data for diamond, the solid line is calculated for $\Theta = 1320$ K (after A. Einstein, Annalen der Physik vol.22, p.186 (1907)).

Its high temperature limit $C_v = 3Nk_B = 25\,\text{J.K}^{-1}\text{mol}^{-1}$ is independent of ω, thus universal : this is the Dulong and Petit law (1819), valid for metals at room temperature.

The low temperature limit $C_v \to 0$ for $T \to 0$ is in agreement with experiment. However, the above law of variation of C_v is not experimentally verified. The Einstein model is improved by introducing a repartition of characteristic frequencies in a solid instead of a single ω : this is the Debye model (1913), in which C_v is proportional to T^3 at very low temperatures. The agreement with experiment then becomes excellent for insulators, whereas for metals an additional term is due to the mobile electron contribution to the specific heat (see § (7.2.2)).

2.5 Exchange of Both Energy and Particles : "Grand Canonical Ensemble"

Here we consider two coupled systems A and A', separated by a porous partition through which energy and particles can be exchanged, the combined system A_0 being isolated from its surroundings (Fig. 2.6). An example of such a situation is water molecules under two phases, liquid and gas, contained in

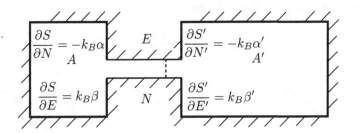

FIG. 2.6 : The two systems A and A' exchange energy and particles.

a thermally insulated can of fixed volume, the water molecules belonging to either phase. One can also study hydrogen molecules as a free gas in equilibrium with hydrogen molecules adsorbed on a solid palladium catalyst surface together with the total system being thermally insulated and the total number of molecules fixed.

We have to state that the entropy is maximum at equilibrium, following an approach very similar to the one of 2.3.1.

2.5.1   Equilibrium Condition : Equality of Both Temperatures and Chemical Potentials

One writes that the maximum of entropy for the combined system $A_0 = A+A'$ takes place at constant total energy and fixed total number of particles, i.e.,

$$S_{A_0}(E_0, N_0) = S_A(E, N) + S_{A'}(E', N') \text{ maximum} \qquad (2.52)$$

under the two constraints

$$\begin{cases} E_0 = E + E' \\ N_0 = N + N' \end{cases} \qquad (2.53)$$

Let us express the condition of entropy maximum of A_0 :

$$S_{A_0}(E_0, N_0) = S_A(E, N) + S_{A'}(E_0 - E, N_0 - N) \qquad (2.54)$$

With respect to an energy exchange between the two systems :

$$\left(\frac{\partial S_{A_0}(E, N)}{\partial E}\right)_N = 0 ,$$

i.e., $\left(\dfrac{\partial S_{A_0}(E, N)}{\partial E}\right)_N = \left(\dfrac{\partial S_A(E, N)}{\partial E}\right)_N + \left(\dfrac{\partial S_{A'}(E_0 - E, N_0 - N)}{\partial E}\right)_N = 0$

$$(2.55)$$

With respect to a particles exchange :

$$\left(\frac{\partial S_{A_0}(E, N)}{\partial N}\right)_E = 0$$

i.e., $\left(\dfrac{\partial S_{A_0}(E, N)}{\partial N}\right)_E = \left(\dfrac{\partial S_A(E, N)}{\partial N}\right)_E + \left(\dfrac{\partial S_{A'}(E_0 - E, N_0 - N)}{\partial N}\right)_E = 0.$

$$(2.56)$$

The thermal equilibrium condition (2.55) again provides the results of §2.3.1, that is, $\beta = \beta'$. The condition (2.56) on N implies that

$$\left(\frac{\partial S_A(E, N)}{\partial N}\right)_E = -\left(\frac{\partial S_{A'}(E_0 - E, N_0 - N)}{\partial N}\right)_E = \left(\frac{\partial S_{A'}(E', N')}{\partial N'}\right)_{E'}$$

$$(2.57)$$

The second equilibrium condition is the equality, in the systems A and A' in contact, of the partial derivatives of the entropy versus the number of particles, all the other parameters being kept constant. This derivative will be written

$$\left(\frac{\partial S}{\partial N}\right)_{\Omega, E} = -k_B \alpha \qquad (2.58)$$

where k_B is the Boltzmann constant. It will be seen in §3.5.1 that the parameter α is related to the chemical potential μ through $\alpha = \beta\mu = \mu/k_B T$.

The equilibrium condition between systems A and A' thus simultaneously implies

$$\begin{cases} \beta = \beta' \text{ , i.e., } T = T' \\ \\ \alpha = \alpha' \text{ , i.e., } \dfrac{\mu}{T} = \dfrac{\mu'}{T'} \end{cases} \qquad (2.59)$$

that is, the equality of *both* the temperatures and the chemical potentials in both systems. Thus in equilibrium the chemical potential of hydrogen is the same, whether in gas phase or adsorbed on the catalyst, and the temperatures are the same in both phases.

2.5.2 Heat Reservoir and Particles Reservoir ; Grand Canonical Probability and Partition Function

Similarly to the approach of §2.4, one now considers that the system A under study is macroscopic, but much smaller that the system A' (Fig. 2.7). Consequently, the parameters α' and β' of system A' are almost the same as those of the combined system A_0, and are very little modified when system A varies in energy or in number of particles : with respect to A the system A' behaves like a heat reservoir and a particle reservoir, it dictates both its temperature $\beta' = 1/k_B T'$ and its chemical potential $\alpha' = \mu/k_B T'$ to A.

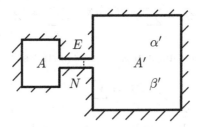

FIG. 2.7: The system A is in contact with the heat reservoir and particles reservoir A', which dictates its parameters α' and β' to A.

This situation in Statistical Physics, in which the system under study A is coupled to a heat reservoir that dictates its value of β, i.e., its temperature, and to a particle reservoir that gives the value of α, is called the "grand canonical ensemble."

2.5.3 Grand Canonical Probability Law and Partition Function

Now both the energy E_i and the number of particles N_i of the considered system enter into the Boltzmann-type probability p_i. Indeed, consider the probability p_i to reach *a particular microstate* i of A, of energy E_i and particle number N_i. It is related to the ability for the *reservoir* to realize this physical situation at given energy and number of particles, that is, to the number $W_{\text{res}}(E_0 - E_i, N_0 - N_i)$ of equally likely microstates, and thus to the reservoir entropy, through the Boltzmann entropy relation. Now

$$S_{\text{res}}(E_0 - E_i, N_0 - N_i) = S_{A_0}(E_0, N_0) - E_i \left(\frac{\partial S_{\text{res}}}{\partial E}\right)_{E_0, N_0}$$
$$- N_i \left(\frac{\partial S_{\text{res}}}{\partial E}\right)_{E_0, N_0} + \ldots \qquad (2.60)$$

thus

$$\ln W_{\text{res}}(E_0 - E_i, N_0 - N_i) = \ln W_{\text{res}}(E_0, N_0) - E_i \beta_0 + N_i \alpha_0 + \ldots \qquad (2.61)$$

The probability of occurrence of this microstate is proportional to W_{res}, i.e.,

$$p(E_i, N_i) = \frac{1}{Z_G} \exp(-\beta_0 E_i + \alpha_0 N_i) \qquad (2.62)$$

which defines the grand partition function Z_G, the quantity which norms the probabilities. It is now a sum over all energies and numbers of particles that can be achieved; this sum can also be regrouped according to the different numbers of particles :

$$Z_G = \sum_i e^{-\beta E_i + \alpha N_i} = \sum_N e^{\alpha N} Z_N \qquad (2.63)$$

The canonical partition function Z_N for N particles, as defined in (2.38),

$$Z_N = \sum_i e^{-\beta E_i} \qquad (2.64)$$

thus appears in Z_G with a weight equal to $\exp \alpha N$.

(Note that in (2.63) the sum is performed over all microstates i with an *arbitrary number* of particles, whereas in (2.64) the sum is only over the states with N particles, with N *fixed*.)

2.5.4   Average Values

Using $p(E_i, N_i)$ one calculates the average energy and the average number of particles of system A : this is conveniently performed from the partial derivatives of Z_G. The approach is the same as in (2.39) :

$$\langle E \rangle = \sum_i E_i p(E_i, N_i) = \sum_i E_i \frac{\exp(-\beta E_i + \alpha N_i)}{Z_G} = -\frac{1}{Z_G} \frac{\partial Z_G}{\partial \beta} = -\frac{\partial \ln Z_G}{\partial \beta}$$
$$(2.65)$$

$$\langle N \rangle = \sum_i N_i p(E_i, N_i) = \sum_i N_i \frac{\exp(-\beta E_i + \alpha N_i)}{Z_G} = +\frac{1}{Z_G} \frac{\partial Z_G}{\partial \alpha} = \frac{\partial \ln Z_G}{\partial \alpha}$$
$$(2.66)$$

To determine the energy fluctuations around its average value, one proceeds in a way similar to the calculation in the canonical ensemble (eqs. (2.41) and (2.42)). Let us calculate the fluctuation of the particle number in system A :

$$\langle N^2 \rangle = \sum_i N_i^2 p(E_i, N_i) = \frac{1}{Z_G} \frac{\partial^2 Z_G}{\partial \alpha^2} \qquad (2.67)$$

$$(\Delta N)^2 = \langle N^2 \rangle - \langle N \rangle^2 = \frac{1}{Z_G} \frac{\partial^2 Z_G}{\partial \alpha^2} - \left(\frac{1}{Z_G} \frac{\partial Z_G}{\partial \alpha} \right)^2 \qquad (2.68)$$

$$= \frac{\partial}{\partial \alpha} \left(\frac{1}{Z_G} \frac{\partial Z_G}{\partial \alpha} \right) = \frac{\partial^2 \ln Z_G}{\partial \alpha^2}$$

Since $\ln Z_G$ is extensive as it increases like $\langle E \rangle$ or $\langle N \rangle$ (see (2.65) and (2.66)), $(\Delta N)^2$ varies proportionaly with the size of the system, or with its average number of particles. Consequently, $\Delta N/N$ is of the order $\langle N \rangle^{-1/2}$, it is thus very small for a macroscopic system.

Thus by differentiating Z_G one obtains the average physical parameters $\langle E \rangle$ and $\langle N \rangle$ and their fluctuations around these averages : in fact Z_G, which expresses the probabilities, contains the complete information on the system.

2.5.5 Grand Canonical Entropy

Using the Gibbs definition

$$S = -k_B \sum_i p_i \ln p_i \tag{2.69}$$

one calculates the entropy in the grand canonical ensemble. Here

$$p_i = \frac{1}{Z_G} e^{-\beta E_i + \alpha N_i} \ , \quad \sum_i p_i = 1 \tag{2.70}$$

Consequently,

$$S = -k_B \sum_i p_i (-\ln Z_G - \beta E_i + \alpha N_i) \tag{2.71}$$

$$S = k_B (\ln Z_G + \beta \langle E \rangle - \alpha \langle N \rangle) = k_B \ln Z_G + k_B \beta U - k_B \alpha \langle N \rangle \tag{2.72}$$

In the present definition of the grand canonical ensemble, we have considered a macroscopic system in contact with a heat reservoir and a particle reservoir. The grand canonical ensemble is also used when the system under study is specified by the average values of both its energy and its particle number and one then needs to determine β and $\alpha = \beta\mu$: the argument is the extension of the one in § 2.4.2. We will use the grand canonical ensemble in particular for the study of the Quantum Statistics : the Pauli principle introduces limitations to the occupation of the quantum states, which will be more conveniently expressed in this ensemble.

2.6 Other Statistical Ensembles

As you imagine, one can generalize to various physical situations the approach consisting of considering two weakly coupled systems A and A', the combined system being isolated, i.e., the total parameters being fixed. In particular one can study systems A and A' exchanging both energy and volume, the total volume $\Omega_0 = \Omega + \Omega'$ being fixed.

The equilibrium condition now implies that the partial derivatives of the entropy

$$\left(\frac{\partial S}{\partial E}\right)_{N,\Omega} = k_B\beta \quad \text{and} \quad \left(\frac{\partial S}{\partial \Omega}\right)_{N,E} = k_B\gamma \qquad (2.73)$$

take the same value in both systems. Thermodynamics will show, in § 3.5.1, that $k_B\gamma = P/T$ where P is the pressure, so that the new equilibrium condition will be

$$\begin{cases} T = T' \\ P = P' \end{cases} \qquad (2.74)$$

that is, in equilibrium the systems A and A' will have the same temperature and the same pressure. If A' is much larger than A, it will give its pressure to A.

The chemistry experiments performed in isothermal and isobar conditions (for example, at room temperature and under the atmospheric pressure) are in such a condition. Then

$$p(E_i, \Omega_i) = \frac{1}{Z_{T,P}} \exp(-\beta E_i - \gamma\Omega_i) = \frac{1}{Z_{T,P}} \exp\left(-\frac{E_i}{k_B T} - \frac{P\Omega_i}{k_B T}\right) \qquad (2.75)$$

i represents the microstate in which the system has the energy E_i and occupies the volume Ω_i. A new partition function $Z_{T,P}$ is defined by

$$Z_{T,P} = \sum_i \exp(-\beta E_i - \gamma\Omega_i) \qquad (2.76)$$

(In fact the sum over the volume, a continuous variable, is rather an integral.) The entropy is deduced :

$$S = -k_B \sum_i p(E_i, \Omega_i)(-\ln Z_{T,P} - \beta E_i - \gamma\Omega_i) \qquad (2.77)$$

$$S = k_B(\ln Z_{T,P} + \beta\langle E \rangle + \gamma\langle \Omega \rangle) \qquad (2.78)$$

Summary of Chapter 2

In this chapter the general method for solving problems in Statistical Physics is presented.

– First one needs to determine the quantum states of the system. In the cases we will consider, the system hamiltonian can be decomposed into a sum of similar one-particle terms

$$\hat{H} = \sum_i \hat{h}_i$$

Until chapter 4, the particles will be assumed to be distinguishable.

– Only in a second step does one choose a statistical ensemble adapted to the physical conditions of the problem (or the most convenient to solve the problem, since for macroscopic systems one cannot measure the fluctuations resulting from the choice of a specific ensemble rather than another one). One expresses that the entropy is maximum in equilibrium, under the constraints set by the physical conditions.

According to the chosen ensemble, a mathematical quantity deduced from the probabilities allows one to obtain the macroscopic physical parameters that can be measured in experiments :

– *the microcanonical ensemble* is used to treat the case of an isolated system : the number of microstates $W(E)$ all producing the same macrostate of energy in the range between E and $E + \delta E$, allows the entropy to be obtained

$$S = k_B \ln W(E)$$

– *the canonical ensemble* is adapted to the case of a system with a fixed number of particles N, in contact with a heat reservoir at temperature T.

The probability to be in a microstate of energy E_i is then given by the Boltzmann factor :

$$p_i = \frac{e^{-\beta E_i}}{Z_N} \text{ , with } \beta = \frac{1}{k_B T}$$

The canonical partition function for N particles is defined by

$$Z_N = \sum_i e^{-\beta E_i}$$

It is related to the average energy through

$$\langle E \rangle = -\frac{1}{Z}\frac{\partial Z}{\partial \beta} = -\frac{\partial \ln Z}{\partial \beta}$$

– the *grand canonical ensemble* corresponds to the situation of a system in contact with a heat reservoir and a particle reservoir. The grand canonical partition function is introduced :

$$Z_G = \sum_N e^{\alpha N} Z_N$$

The probability to be in a microstate of both energy E_i and number of particles N_i is given by

$$p(E_i, N_i) = \frac{1}{Z_G}\exp(-\beta E_i + \alpha N_i)$$

The average energy and the average number of particles are then deduced :

$$\langle E \rangle = -\frac{1}{Z_G}\frac{\partial Z_G}{\partial \beta} = -\frac{\partial \ln Z_G}{\partial \beta} \ , \ \langle N \rangle = \frac{1}{Z_G}\frac{\partial Z_G}{\partial \alpha} = \frac{\partial \ln Z_G}{\partial \alpha}$$

Appendix 2.1

Lagrange Multipliers

One looks for the extremum of a function with several variables, satisfying constraints. Here we just treat the example of the entropy maximum with a fixed average value of the energy, i.e.,

$$S = -k_B \sum_i p_i \ln p_i \quad \text{maximum, under the constraints}$$

$$\begin{cases} \sum_i p_i = 1 \\ \sum_i p_i E_i = \langle E \rangle \end{cases} \tag{2.79}$$

The process consists in introducing constants, the Lagrange multipliers, here λ and β, the values of which will be determined at the end of the calculation. Therefore one is looking for the maximum of the auxiliary function :

$$F(p_1, \ldots, p_i, \ldots) = -\sum_i p_i \ln p_i + \lambda(\sum_i p_i - 1) - \beta(\sum_i E_i p_i - \langle E \rangle) \tag{2.80}$$

which expresses the above conditions, with respect to the variables p_i. This is written

$$\frac{\partial F}{\partial p_i} = -(\ln p_i + 1) + \lambda - \beta E_i = 0 \tag{2.81}$$

The solution is

$$\ln p_i = -\beta E_i + \lambda - 1 \tag{2.82}$$

Consequently,

$$p_i = (e^{\lambda - 1}) \, e^{-\beta E_i} \tag{2.83}$$

The constants λ and β are then obtained by expressing the constraints :

$$\bullet \quad \sum_i p_i = 1 \ , \quad \text{whence}$$

$$e^{\lambda-1} \sum_i e^{-\beta E_i} = 1 \tag{2.84}$$

$$e^{\lambda-1} = \frac{1}{\displaystyle\sum_i e^{-\beta E_i}} = \frac{1}{Z} \tag{2.85}$$

$$\bullet \quad \sum_i p_i E_i = \langle E \rangle \ , \quad \text{i.e.,} \tag{2.86}$$

$$\frac{1}{Z} \sum_i E_i e^{-\beta E_i} = \langle E \rangle \tag{2.87}$$

which is indeed expression (2.36).

Note that, if the only constraint is $\Sigma_i p_i = 1$, the choice of the p_i maximizing S indeed consists in taking all the p_i equal; this is what is done for an isolated system using the Boltzmann formula.

Chapter 3

Thermodynamics and Statistical Physics in Equilibrium

In Chapter 1 we stated the basic principles of Statistical Physics, which rely on a detailed microscopic description and on the hypothesis, for an isolated system, of the equiprobability of the different microscopic occurrences of a given macroscopic state; the concept of statistical entropy was also introduced. In Chapter 2 the study was extended to other systems in equilibrium, in which the statistical entropy is maximized, consistently with the constraints defined by the physical conditions of the problem. It is now necessary to link this statistical entropy S and the entropy which appears in the second law of Thermodynamics and that you became familiar with in your previous studies. It will be shown in this chapter that these two entropies are identical.

Thus the purpose of this chapter will first be ($\S$ 3.1 to 3.4), mostly in the equilibrium situation, to rediscover the laws of Thermodynamics, which rule the behavior of macroscopic physical properties. This will be done in the framework of the hypotheses and results of Statistical Physics, as obtained in the first two chapters of this book. Then we will be able to use at best the different thermodynamical potentials ($\S$ 3.5), each of them being directly related to a partition function, that is, to a microscopic description. They are convenient tools to solve the Statistical Physics problems.

We are going to state the successive laws of Thermodynamics and to see how each one is interpreted in the framework of Statistical Physics. The example chosen in Thermodynamics will be the ideal gas, the properties of which you studied in previous courses. Its study in Statistical Physics will be presented in the next chapter 4.

3.1   Zeroth Law of Thermodynamics

According to this law, *if two bodies at the same temperature T are brought into thermal contact, i.e. can exchange energy, they remain in equilibrium.*

In Statistical Physics, we saw in (2.3.1) that if two systems are free to exchange energy, the combined system being isolated, the *equilibrium* condition is that the partial derivative of the statistical entropy versus energy should take the same value in each system :

$$\frac{\partial S}{\partial E} = k_B \beta, \ \frac{\partial S'}{\partial E'} = k_B \beta'; \beta = \beta' \tag{3.1}$$

It was suggested that this partial derivative of the entropy is related to its temperature, as will be more precisely shown below in § 3.3.

3.2   First Law of Thermodynamics

Its statement is as follows :

Consider an infinitesimal process of a system with a fixed number of particles, exchanging an infinitesimal work δW and an infinitesimal amount of heat δQ with its surroundings. Although both δW and δQ depend on the sequence of states followed by the system, the sum

$$dU = \delta W + \delta Q \tag{3.2}$$

only depends on the initial state and the final state, i.e., it is independent of the process followed. This states that dU is the differential of a state function U, which is the internal energy of the system.

A property related to this first law is that the internal energy of an isolated system remains constant, as no exchange of work or heat is possible :

$$dU = 0 \tag{3.3}$$

In this Statistical Physics course, we introduced the energy of a macroscopic system : in the microcanonical ensemble it is given exactly ; in the canonical or grand canonical ensemble its value is given by the average

$$\langle E \rangle = U = \sum_i p_i E_i \tag{3.4}$$

where p_i is the probability of occurrence of the state of energy E_i. The differential of the internal energy

$$dU = \sum_i p_i dE_i + \sum_i E_i dp_i \qquad (3.5)$$

consists in two terms that we are going to identify with the two contributions in dU, i.e., δW and δQ. For this purpose, two particular processes will first be analyzed, in which either work (§ 3.2.1) or heat (§ 3.2.2) is exchanged with the surroundings. Then the case of a quasi-static general process will be considered (§ 3.2.3).

3.2.1 Work

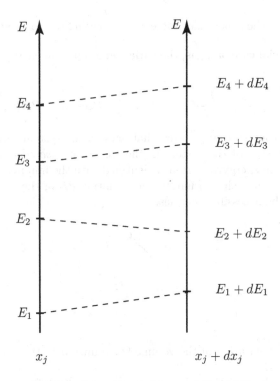

FIG. 3.1: Modification of the energies of the considered system when the external parameter varies from x_j to $x_j + dx_j$.

The quantum states of a system are the solutions of a hamiltonian which depends of external parameters x_j characteristic of the system : for example, you learnt in Quantum Mechanics, and will see again in chapter 6, that a gas molecule, free and confined within a fixed volume Ω, assumes for energy eigenvalues $\hbar^2 k^2/2m$, where the allowed values of $\vec{k}$ depend of the container volume like $\Omega^{-1/3}$; as for the probabilities of realizing these eigenstates, they are determined by the postulates of Statistical Physics.

Mechanical work will influence these external parameters : for example, the motion of a piston modifies the volume of the system, thus shifting its quantized states. More generally, if only the external parameters of the hamiltonian are modified from x_j to $x_j + dx_j$, the energy of the eigenstates is changed and a particular eigenvalue E_i of the hamiltonian becomes $E_i + dE_i$ in such a process (Fig. 3.1).

The variation of the average energy of the system is then

$$\frac{\partial U}{\partial x_j}dx_j = -X_j dx_j \tag{3.6}$$

where X_j is the generalized force corresponding to the external parameter x_j.

In the special case of a gas, the variation of internal energy in a dilatation dL is equal to

$$\frac{\partial U}{\partial L}dL = -FdL = -P\mathcal{S}dL = -Pd\Omega \tag{3.7}$$

Here P is the external pressure that acts on the system surface $\mathcal{S}$; it is equal to the pressure of the system in the case of a *reversible* process (Fig. 3.2). This internal energy variation is identified with the infinitesimal work received by the gas. In such a situation the variation dL of the characteristic length modifies the accessible energies.

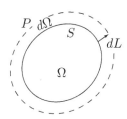

FIG. 3.2 : Dilatation of the volume Ω submitted to the external pressure P.

In general the variation of average energy resulting from the variation of the external parameters is identified with the infinitesimal work

$$\delta W = -\sum_j X_j dx_j \tag{3.8}$$

and corresponds to the term $\sum_i p_i dE_i$

3.2.2 Heat

Assume that *energy is brought* into the system, while *no work is delivered* : the external parameters of the hamiltonian are unchanged, the energy levels E_i remain identical. The average energy of the system increases, now due to higher probabilities of occurrence of the system states of higher energy. The probabilities p_i are modified and now the variation of internal energy is identified with the infinitesimal heat absorbed :

$$\delta Q = \sum_i E_i dp_i \qquad (3.9)$$

A way to confirm this identification of $\sum_i E_i dp_i$ is to consider a very slow process in which no heat is exchanged with the surroundings, that is, the system is thermally insulated : this is an *adiabatic** process. In Thermodynamics, such a process is done at constant entropy. In Statistical Physics, it can also be shown that the statistical entropy is conserved in this process :

$$S = -k_B \sum_i p_i \ln p_i = \text{constant} \qquad (3.10)$$

Consequently,

$$\sum_i dp_i \ln p_i + \sum_i p_i \frac{dp_i}{p_i} = 0 \qquad (3.11)$$

Now as $\sum_i p_i = 1$, one has $\sum_i dp_i = 0$, whence

$$\sum_i \ln p_i dp_i = 0 \qquad (3.12)$$

One replaces $\ln p_i$ by its expression in the canonical ensemble

$$\ln p_i = -\beta E_i - \ln Z \qquad (3.13)$$

and deduces

$$\sum_i E_i dp_i = 0 \qquad (3.14)$$

which is indeed consistent with the fact that the heat δQ exchanged with the surroundings is zero in an adiabatic process.

3.2.3 Quasi-Static General Process

Consider an infinitesimal *quasi-static** process of a system, that is, slow enough to consist in a continuous *and reversible** sequence of equilibrium states.

In the differential dU one still identifies the work with the term arising from the variation of the hamiltonian external parameters in this process. The complementary contribution to dU is identified with δQ, the heat exchanged with the surroundings in the process.

3.3   Second Law of Thermodynamics

Let us recall its statement :

Let us consider a system with a fixed number N of particles, in thermal contact with a heat source at temperature T. The system exchanges an amount of heat δQ with this source in an infinitesimal quasi-static process. Then the entropy variation dS_{thermo} of the considered system is equal to

$$dS_{\text{thermo}} = \frac{\delta Q}{T} \tag{3.15}$$

Although δQ depends on the particular path followed, the extensive quantity dS_{thermo} is the differential of a state function, which thus only depends on the initial and final states of the considered process.

Besides, in a spontaneous, i.e. without any contribution from the surroundings, and irreversible process, the entropy variation dS_{thermo} is positive or zero.

The general definition of the entropy in Statistical Physics is that of Gibbs :

$$S = -k_B \sum_i p_i \ln p_i \tag{3.16}$$

Let us first examine the interpretation of an irreversible process in Statistical Physics. An example is the free expansion of a gas into vacuum, the so-called "Joule-Gay-Lussac expansion" : a N-molecule gas was occupying a volume Ω (Fig. 3.3a), the partition is removed, then the gas occupies the previously empty volume Ω' (Fig. 3.3b). The combined system is isolated and removing the partition requires a negligible amount of work. The accessible volume becomes $\Omega + \Omega'$ and it is extremely unlikely that the molecules again gather in the sole volume Ω. In such a process, which is irreversible in Thermodynamics, the accessible volume for the particles has increased, that is, the constraint on their positions has been removed; the number of cells of the one-particle phase space $(\vec{r}, \vec{p})$ has increased with the accessible volume in this process.

Let us now consider an infinitesimal reversible process, corresponding to a mere modification of the probabilities of occurrence of the same accessible

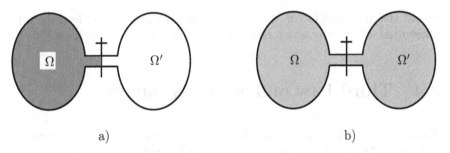

a)   b)

Fig. 3.3 : Joule-Gay-Lussac expansion into vacuum.

microstates. dS is obtained by differentiating (3.16) :

$$dS = -k_B \left(\sum_i \ln p_i dp_i + \sum_i \frac{p_i}{p_i} dp_i \right)$$

$$= -k_B \sum_i \ln p_i dp_i \qquad (3.17)$$

since, as probabilities are normalized, $\sum_i dp_i = 0$.

To be more precise, let us work in the canonical ensemble, in which the system is submitted to this process at constant number of particles, remaining at every time t in thermal equilibrium with a heat reservoir at the temperature $T(t)$. The probability to realize the microstate of energy E_i is equal to (see (2.37)) :

$$p_i(t) = \frac{e^{-\beta(t)E_i(t)}}{Z}, \text{i.e., } \ln p_i(t) = -\beta(t)E_i(t) - \ln Z(t) \qquad (3.18)$$

Then

$$dS = k_B \beta(t) \sum_i E_i(t) dp_i \qquad (3.19)$$

From the above interpretation of the infinitesimal heat δQ, (3.19) can also be written

$$dS = k_B \beta(t) \delta Q \qquad (3.20)$$

This expression is to be compared to (3.15). The *statistical* entropy defined by (3.16) and the *thermodynamical* entropy (3.15) can be identified under the condition that $k_B \beta$, defined by (3.1) as the partial derivative of the statistical entropy S versus energy, is identical to $\frac{1}{T}$, the reciprocal of the thermodynamical temperature :

$$k_B \beta \equiv \frac{1}{T} \qquad (3.21)$$

Indeed the integrand factor, which allows one to transform δQ into the total differential dS, is unique to a constant, which justifies this identification.

3.4   Third Law of Thermodynamics

The second law allows one to calculate entropy *variations*. From the third law, or Walther Nernst (1864-1941)'s law, the *absolute value* of S can be obtained :

When the absolute temperature T of a system tends to zero, its entropy tends to zero.

From a microscopic point of view, when the temperature vanishes, the particles gather into the accessible states with the lowest energies.

If the minimum energy corresponds to a single state E_0, the probability of occupation of this fundamental state tends to unity while those for the other states vanish. Then the microscopic description becomes certain and the entropy vanishes : for example, when the temperature tends to zero in a monoatomic ideal gas, one expects the molecules to be in a state of kinetic energy as small as possible (yet different from zero, because of the Heisenberg uncertainty principle) ; in the case of electrons in a solid, as will be studied in chapter 7, the fundamental state is nondegenerate too.

Let us now consider a material which has two equivalent states at low temperature, with the same minimal energy, for example, two allotropic crystalline species, each occurring with the probability $1/2$. When T tends to zero, the corresponding entropy tends to

$$S_0 = k_B \ln 2 \qquad (3.22)$$

This quantity is different from zero, yet it is not extensive so that S_0/N becomes negligible. This is very different from the situation at finite T where S/N is finite and S extensive.

Thus the statement of the Nernst law consistent with Statistical Physics is that *S/N tends to zero for N large when the temperature tends to the absolute zero.*

Notice that, in the case of the paramagnetic solid at zero external magnetic field and very low temperature, one would expect to find $S/N = k_B \ln 2$, in contradiction with the Nernst law. In fact, this spin system cannot be strictly in zero magnetic field because of the nuclear magnetic interactions that are always present (but are not very efficient at higher temperature where other phenomena imply much larger amounts of energy) : under the effect of these nuclear interactions the system is no longer at zero magnetic field and becomes ordered when T tends to zero.

The present sections have put in correspondence the laws of Thermodynamics and the statistical approach of chapters 1 and 2. Their conclusion is that the description based on Statistical Physics and its postulates is indeed equivalent to the three laws of Thermodynamics.

3.5 The Thermodynamical Potentials and their Differentials; the Legendre Transformation

Now that we have identified the statistical entropy to the thermodynamical entropy, we can take advantage of our background in Thermodynamics to introduce the thermodynamical potential adapted to the considered physical situation, and relate it to the statistical description and to the physical parameters. The thermodynamical potential will be a function of the parameters determined by the experimental conditions and in equilibrium it will be extremum.

3.5.1 Isolated System

Here the entropy S is the thermodynamical potential to consider. During a spontaneous evolution toward equilibrium, S increases and reaches its maximum in equilibrium.

In the microcanonical description, the entropy is related to the number W of microstates of energies lying in the range between E and $E + dE$ by

$$S = k_B \ln W \tag{3.23}$$

dS and dU are linked through the differential expressions of the first and second laws of Thermodynamics :

$$dS = \frac{\delta Q}{T} = \frac{dU}{T} - \frac{\delta W}{T} \tag{3.24}$$

so that

$$\frac{1}{T} = \left(\frac{\partial S}{\partial U}\right)_{N,\Omega} \tag{3.25}$$

In the case of the ideal gas where

$$\delta W = -P d\Omega \tag{3.26}$$

$$dS = \frac{dU}{T} + \frac{P d\Omega}{T} \tag{3.27}$$

The differential expression (3.27) shows that S, calculated for a fixed number N of particles, is a function of the variables U and Ω, both extensive. The pressure is deduced from the partial derivative of S versus the volume :

$$\frac{P}{T} = \left(\frac{\partial S}{\partial \Omega}\right)_{N,U} = \left(\frac{k_B \partial \ln W(U)}{\partial \Omega}\right)_{N,U} \tag{3.28}$$

Note : we suggested in the Appendix 3 of chapter 1 that, in the case of free particles, the number of microstates associated with the energy E was of the form

$$W(E) \propto \Omega^N \chi(E) \tag{3.29}$$

so that

$$\frac{P}{T} = \left(\frac{k_B \partial \ln W(U)}{\partial \Omega}\right)_{N,U} = \frac{k_B N}{\Omega} \tag{3.30}$$

$$P\Omega = N k_B T \tag{3.31}$$

Here we find the ideal gas law, that will be again established in the next chapter.

After having studied a given system, we now consider another similar system, with the same energy U and under the same volume Ω, which *only differs by its number of particles* equal to $N + dN$. Its entropy is now $S(U, \Omega, N) + dS$, and we write

$$dS = -\frac{\mu}{T} dN \tag{3.32}$$

This defines the chemical potential μ, related to the entropy variation of the system when an extra particle is added, at constant internal energy and volume. The chemical potential is an intensive parameter. For a classical ideal gas, if N increases, one expects S to increase, so that μ is negative.

Then, including the dependence in the particles number, the differential of S for the ideal gas is written

$$dS = \frac{dU}{T} + \frac{P d\Omega}{T} - \frac{\mu dN}{T} \tag{3.33}$$

S is a function of extensive variables only; the partial derivative of S versus N is equal to

$$\left(\frac{\partial S}{\partial N}\right)_{\Omega,U} = -\frac{\mu}{T} \tag{3.34}$$

We previously showed in §2.5.1 that if the two systems are exchanging particles, in equilibrium

$$\frac{\partial S}{\partial N} = -k_B \alpha \tag{3.35}$$

takes the same value in both systems.

These two partial derivatives of the entropy versus the number of particles can be identified, so that

$$\alpha = \frac{\mu}{k_B T} \tag{3.36}$$

Besides, from the differential (3.33) of S expressed for the ideal gas one deduces the expression of the differential of its energy U :

$$dU = -Pd\Omega + TdS + \mu dN \tag{3.37}$$

The last term can be interpreted as a "chemical work."

3.5.2 System with a Fixed Number of Particles, in Contact with a Heat Reservoir at Temperature T

We are going to show that the corresponding thermodynamical potential is the Helmholtz free energy F, which is minimum in equilibrium and is related to the N-particle canonical partition function Z_N. Indeed the general definition (3.16) of S applies, with now probabilities p_i equal to

$$p_i = \frac{1}{Z_N} \exp(-\beta E_i) \quad \text{and} \quad \beta = \frac{1}{k_B T} \tag{3.38}$$

Then

$$\sum_i p_i \ln p_i = \sum_i p_i \left(-\frac{E_i}{k_B T}\right) - \ln Z_N$$

$$= -\frac{U}{k_B T} - \ln Z_N \tag{3.39}$$

Consequently, in the canonical ensemble,

$$S = \frac{U}{T} + k_B \ln Z_N \tag{3.40}$$

$$U - TS = -k_B T \ln Z_N \tag{3.41}$$

Now in Thermodynamics the free energy F is defined by

$$F = U - TS \tag{3.42}$$

whence we deduce the relation between the thermodynamical potential F, and the partition function Z_N expressing the Statistical Physics properties :

$$F = -k_B T \ln Z_N \tag{3.43}$$

Let us write the differential of the thermodynamical potential F in the case where the infinitesimal work is given by $\delta W = -\sum_j X_j dx_j$ (see eq. 3.8) :

$$dF = dU - TdS - SdT$$
$$= -\sum_j X_j dx_j + \delta Q + \mu dN - TdS - SdT \tag{3.44}$$

Now $\delta Q = TdS$, so that

$$dF = -\sum_j X_j dx_j - SdT + \mu dN \tag{3.45}$$

The interpretation of (3.45) is that, in a specific infinitesimal reversible process at constant temperature and fixed number of particles, the free energy variation is equal to the work received by the system, whence the name of Helmholtz "free energy," i.e., available energy, for this thermodynamical potential.

For a gas, the infinitesimal work is expressed as a function of pressure, so that

$$dF = -Pd\Omega - SdT + \mu dN \tag{3.46}$$

The entropy, the pressure, and the chemical potential are obtained as partial derivatives F or of $\ln Z_N$:

$$S = -\left(\frac{\partial F}{\partial T}\right)_{N,\Omega} = \left(\frac{\partial (k_B T \ln Z_N)}{\partial T}\right)_{N,\Omega} \tag{3.47}$$

$$P = -\left(\frac{\partial F}{\partial \Omega}\right)_{N,T} = \left(\frac{\partial (k_B T \ln Z_N)}{\partial \Omega}\right)_{N,T} \tag{3.48}$$

$$\mu = \left(\frac{\partial F}{\partial N}\right)_{\Omega,T} = -\left(\frac{\partial (k_B T \ln Z_N)}{\partial N}\right)_{\Omega,T} \tag{3.49}$$

3.5.3   System in Contact with Both a Heat Reservoir at Temperature T and a Particle Reservoir

It was shown in 2.5.1 that, during such an exchange of energy and particles, the parameters taking the same value in equilibrium in both the system and the reservoir are $k_B\beta = 1/T$ and $k_B\alpha = \mu/T$: two Lagrange parameters thus come into the probabilities, which do not only depend on the energy but also on the number of particles. Thus the probability of occurrence of a macroscopic state of energy E_i and number of particles N_i can be written (see § 2.5.3, eqs. (2.62) and (2.63)) :

$$p_i(E_i, N_i) = \frac{1}{Z_G} \exp(-\beta E_i + \alpha N_i) \tag{3.50}$$

so that

$$\sum_i p_i \ln p_i = \sum_i p_i \left(-\frac{E_i}{k_B T} + \alpha N_i - \ln Z_G\right) \tag{3.51}$$

$$= -\frac{U}{k_B T} + \alpha\langle N\rangle - \ln Z_G \tag{3.52}$$

Consequently,

$$S = \frac{U}{T} - \alpha k_B \langle N\rangle + k_B \ln Z_G \tag{3.53}$$

$$U - TS - \alpha k_B T\langle N\rangle = -k_B T \ln Z_G \tag{3.54}$$

where $\langle N\rangle$ is the average number of particles in the system under study.

The thermodynamical potential introduced here is the grand potential A, defined by

$$A = U - TS - \mu\langle N\rangle = F - \mu\langle N\rangle \tag{3.55}$$

It is related to the grand canonical partition function through

$$A = -k_B T \ln Z_G \tag{3.56}$$

The differential of A is deduced from its definition :

$$dA = dF - \mu d\langle N\rangle - \langle N\rangle d\mu \tag{3.57}$$

$$= -\sum_j X_j dx_j - SdT - \langle N\rangle d\mu \tag{3.58}$$

Like the other thermodynamical potentials, A is extensive. The only extensive variable in A is the one in the infinitesimal work. In the case of a gas, it is the volume so that

$$A = -P\Omega \tag{3.59}$$

The entropy S and the average number of particles $\langle N\rangle$ are obtained by calculating the partial derivatives of A.

As a conclusion of this section on the thermodynamical potentials, remember that to a fixed physical situation, corresponding to data on specific external parameters, is associated a particular thermodynamical potential, which is extremum in equilibrium.

3.5.4 Transformation of Legendre ; Other Thermodynamical Potentials

In § 3.5.2 and § 3.5.3 we saw two different physical situations, associated with two thermodynamical potentials with similar differentials, except for a change in one variable. The mathematical process that shifts from $F(N)$ to $A(\mu)$ is called "transformation of Legendre."

The general definition of such a transformation is the following : one goes from a function $\Phi(x_1, x_2, ...)$ to a function

$$\Gamma = \Phi(x_1, x_2, ...) - x_1 y_1, \text{ with } y_1 = \frac{\partial \Phi}{\partial x_1} \tag{3.60}$$

the new function Γ is considered as a function of the new variable y_1. Because of the change of variable and of function, the differentials of Φ and of Γ are respectively :

$$d\Phi = \frac{\partial \Phi}{\partial x_1} dx_1 + \frac{\partial \Phi}{\partial x_2} dx_2 + \ldots \tag{3.61}$$

$$d\Gamma = \frac{\partial \Gamma}{\partial y_1} dy_1 + \frac{\partial \Gamma}{\partial x_2} dx_2 + \ldots \tag{3.62}$$

Now

$$\frac{\partial \Gamma}{\partial y_1} = \frac{\partial \Gamma}{\partial x_1} \cdot \frac{\partial x_1}{\partial y_1} = \left(\frac{\partial \Phi}{\partial x_1} - \frac{\partial \Phi}{\partial x_1} - x_1 \frac{\partial y_1}{\partial x_1} \right) \cdot \frac{\partial x_1}{\partial y_1} = -x_1 \tag{3.63}$$

Consequently,

$$d\Gamma = -x_1 dy_1 + \frac{\partial \Phi}{\partial x_2} dx_2 + \ldots \tag{3.64}$$

The partial derivatives of Γ versus the variables $x_2 \ldots$ are not modified with respect to those of Φ for the same variables.

In this process one has shifted from the variables $(x_1, x_2 \ldots x_n)$ to the variables $(y_1, x_2 \ldots x_n)$. This is indeed the transformation that was performed for example, when shifting from $U(S)$ to $F(T)$, the other variables Ω, N being conserved, or when shifting from $F(N)$ to $A(\mu)$, the variables Ω and T being unchanged.

Other thermodynamical potentials can be defined in a similar way, a specific potential corresponding to each physical situation. For example, a frequent situation in Chemistry is that of experiments performed at *constant temperature and constant pressure* (isothermal and isobaric condition) : the exchanges occur between the studied system A and the larger system A' which acts as an energy and volume reservoir. We saw that the equilibrium condition, expressing the maximum of entropy of the combined system with respect to exchanges of energy and volume, corresponds to the equality of the temperatures and the pressures in both systems. The corresponding thermodynamical potential is the Gibbs free enthalpy G, in which the number of particles is fixed :

$$G = U - TS + P\langle\Omega\rangle$$
$$dG = \langle\Omega\rangle dP - SdT + \mu dN \tag{3.65}$$

This potential is defined from the free energy F

$$F = U - TS$$
$$dF = -Pd\Omega - SdT + \mu dN \tag{3.66}$$

using a transformation of Legendre on the set of variables (volume Ω and pressure P), that is

$$G = F + P\langle\Omega\rangle \tag{3.67}$$

The chemical potential is interpreted here as the variation of the free enthalpy G when one particle is added to the system, at constant pressure and temperature. The free enthalpy, extensive like all thermodynamical potentials, depends of a single extensive variable, so that

$$G = \mu N \tag{3.68}$$

The chemical potential μ is here the free enthalpy per particle.

The free enthalpy is related to the partition function which includes all the possible values of energy and volume of the considered system (see § 2.6) :

$$Z_{T,P} = \sum_i e^{-\beta E_i - \gamma\Omega_i} \tag{3.69}$$

Note 1 : each time a transformation of Legendre is performed, an extensive variable is replaced by an intensive one. Now a thermodynamical potential has to be extensive, so that it should retain at least one extensive variable. Consequently, there is a limit to the number of such tranfomations yielding a new thermodynamical potential.

Note 2 : Appendix 4.1 presents a practical application of thermodynamical potentials to the study of chemical reactions.

Summary of Chapter 3

We have expressed the laws of Thermodynamics in the language of Statistical Physics.

– *Zeroth law* :

For two systems A and A' in thermal equilibrium

$$\frac{1}{T} = k_B \beta = \left(\frac{\partial S}{\partial E} \right)_{N, \Omega} = k_B \beta' = \frac{1}{T'}$$

– *First law* :

The differential of the average energy of a system :

$$dU = d \left(\sum_i p_i E_i \right) = \sum_i p_i dE_i + \sum_i E_i dp_i$$

is split into two terms.

The infinitesimal work is related to the modification of the energy states of the system due to the change of external parameters (volume, etc.)

$$\delta W = \sum_i p_i dE_i$$

The infinitesimal heat exchanged δQ is associated with variations of probabilities of occurrence of the microstates

$$\delta Q = \sum_i E_i dp_i$$

– *Second law* :

We have identified the thermodynamical entropy with the statistical entropy as introduced in the first chapters of this book and have shown that $k_B \beta = \frac{1}{T}$.

In Statistical Physics, an irreversible process corresponds to an increase of the number of accessible microstates for the considered system, and consequently to an increase of disorder and of entropy.

– *Third law* :

The entropy per particle S/N vanishes when the temperature tends to the absolute zero.

– The adapted *thermodynamical potential* depends on the considered physical situation, it is extremum in equilibrium. It is related to the microscopic parameters of Statistical Physics through the partition function ; its partial derivatives give access to the measurable physical parameters. Using a transformation of Legendre, one gets a new thermodynamical potential, with one new variable (see the table next page).

Ensemble	Microcanonical	Canonical	Grand Canonical
Physical Conditions	*isolated* system E to δE	*exchange of energy with a heat reservoir* at fixed N	*exchange of energy with a heat reservoir and of particles with a particle reservoir*
Lagrange Parameter(s)		$\beta = \dfrac{1}{k_B T}$	$\beta = \dfrac{1}{k_B T}$, $\alpha = \beta\mu = \dfrac{\mu}{k_B T}$
Statistical Description	W microscopic configurations at E	$Z_N(\beta) = \displaystyle\sum_{\text{microstates}} e^{-\beta E_n}$	$Z_G(\alpha, \beta) = \displaystyle\sum_N e^{\alpha N} Z_N(\beta)$
Thermodynamical Potential	$S = k_B \ln W$	$F = -k_B T \ln Z_N$ $F = U - TS$	$A = -k_B T \ln Z_G$ $A = U - TS - \mu N$
Thermodynamical Parameters	$dS = \dfrac{P}{T} d\Omega + \dfrac{dU}{T} - \dfrac{\mu}{T} dN$ $\dfrac{1}{T} = \left(\dfrac{\partial S}{\partial U}\right)_{\Omega, N} \cdots$	$dF = -P d\Omega - S dT + \mu dN$ $S = -\left(\dfrac{\partial F}{\partial T}\right)_{\Omega, N} \cdots$	$dA = -P d\Omega - S dT - N d\mu$ $\langle N \rangle = -\left(\dfrac{\partial A}{\partial \mu}\right)_{\Omega, T}$ *(average value)*

Chapter 4

The Ideal Gas

4.1 Introduction

After having stated the general methods in Statistical Physics and bridged its concepts with your previous background in Thermodynamics, we are now going to apply our know-how to the molecules of a gas, with no mutual interaction. The properties of such a gas, called "ideal gas" ("perfect gas" in French) are discussed in Thermodynamics courses. Here the contribution of Statistical Physics will be evidenced : starting from a microscopic description in *Classical Mechanics*, it allows to explain the observed macroscopic properties.

A few notions of kinetic theory of gases will first be introduced in § 4.2 and § 4.3 we will discuss to what extent *classical Statistical Physics*, as used up to this point in this book, can apply to free particles. This chapter is "doubly classical" since, like in the three preceding chapters, Classical Statistics is used, but now for particles with an evolution described by Classical Mechanics : both the states themselves and their occurrence are classical. The canonical ensemble is chosen in § 4.4 to solve the statistical properties of the ideal gas and physical consequences are deduced ; in the Appendices, all that we learnt in this chapter and the preceding ones is applied in a practical problem : the interpretation of chemical reactions in gas phase is done in Thermodynamics (Appendix 4.1) and in Statistical Physics (Appendix 4.2).

4.2   Kinetic Approach

This section, which has no relation with Quantum Mechanics or Statistical Physics in equilibrium, will allow us to obtain insight into the simplest physical parameters describing the motion of molecules in a gas, and to introduce some orders of magnitude. If necessary, we will refer to the Thermodynamics courses on the ideal gas.

4.2.1   Scattering Cross Section, Mean Free Path

Consider a can filled with a gas, in the standard conditions of temperature and pressure ($T = 273$ K, $P = 1$ atmosphere). Each molecule has a radius r, defined by the average extension of the valence orbitals, i.e., a fraction of a nanometer. If in its motion a molecule comes too close to another one, there will be a collision. One defines the *scattering cross section* which is the circle of radius r around a molecule, inside which no other molecule can penetrate.

FIG. 4.1: Definitions of the scattering cross section and of the mean free path.

The mean distance between two collisions l, or *mean free path*, is deduced from the mean volume per particle Ω/N and the scattering cross section by

$$l(\pi r^2) = \frac{\Omega}{N} = d^3 \tag{4.1}$$

where d is the average distance between particles (Fig. 4.1). From l, one deduces the mean time τ between two collisions, or collision time, by writing

$$l = \nu\tau \tag{4.2}$$

where ν is the mean velocity at temperature T, of the order of $\sqrt{k_B T/m}$ (see the justification in § 4.4.2).

For nitrogen molecules with $r = 0.42$ nm, at 273 K and under a pressure of one atmosphere one finds $l \approx 70$ nm, $\nu \approx 300 \, \text{m.sec}^{-1}$, $\tau \approx 2 \times 10^{-10}$ sec. The length l is to be compared with the average distance between particles $d = 3.3$ nm. These values show that in the gas collisions are very frequent at the time scale of a measure, which ensures that equilibrium is rapidly reached after a perturbation of the gas; at the molecular scale distances covered between collisions are very large (of the order of a hundred times the size of a molecule), thus the motion of each molecule mostly occurs as if it was alone, that is, free.

4.2.2 Kinetic Calculation of the Pressure

Let us now recall the calculation, from kinetic arguments, of the pressure of a gas, i.e., of the force exerted on a wall of unit surface by the molecules. The aim is to evaluate the total momentum transferred to this wall by the molecules impinging on it during the time interval Δt : indeed in this chapter Classical Mechanics does apply, so that the total force on the wall is given by the Fundamental Equation of Dynamics :

$$\vec{F}_{\text{total}} = \sum_i \frac{\Delta \vec{p}_i}{\Delta t} \tag{4.3}$$

Here the momentum brought by particle i is $\Delta \vec{p}_i$. The resulting force is normal since the wall is at rest, it is related to the pressure P and to the considered wall surface $\mathcal{S}$ through

$$|\vec{F}_{\text{total}}| = P\mathcal{S} \tag{4.4}$$

One will now evaluate the total momentum transferred to the surface, which is the difference between the total momentum of the particles which reach the wall and that of the particles which leave it (Fig. 4.2).

The incident particles, with $p_z > 0$, bring a total momentum $\sum_i \vec{p}_i f(\vec{r}, \vec{p}_i, t)$ to the wall, chosen as xOy plane.

The probability density *in equilibrium* $f(\vec{r}, \vec{p}_i, t)$ is assumed to be stationary, homogeneous, and isotropic in space (hypotheses stated by James Clerk Maxwell in 1859). It thus reduces to an *even* function $f(\vec{p})$ versus the three components of the momentum and satisfies to

$$\int f(\vec{p}) d^3\vec{p} = 1 \tag{4.5}$$

Consider the particles, of momentum equal to $\vec{p}$ to $d^3\vec{p}$, incident on the surface $\mathcal{S}$ during the time interval Δt. They are inside an oblique cylinder along the

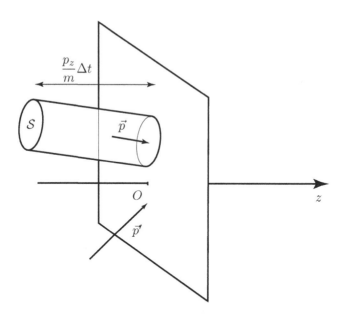

FIG. 4.2 : Collisions of molecules on a wall.

direction of $\vec{p}$, of basis $\mathcal{S}$, height $\dfrac{p_z}{m}\Delta t$ and volume $\dfrac{\mathcal{S}p_z}{m}\Delta t$ (the direction z is chosen along the normal to the surface). For a particles density $\rho = N/\Omega$, the number of incident particles is

$$\frac{\mathcal{S}p_z}{m}\Delta t\, \rho f(\vec{p})d^3\vec{p} \tag{4.6}$$

Each of them transfers its momentum $\vec{p}$ to the wall. The total momentum brought by the incident particles is thus equal to

$$\sum_{\text{inc}}\Delta\vec{p}_i = \frac{\mathcal{S}}{m}\Delta t\rho \int_{p_z>0} p_z\vec{p}f(\vec{p})d^3\vec{p} \tag{4.7}$$

This sum is restricted to the particles *traveling toward the surface*, of momentum $p_z > 0$.

A similar calculation leads to the total momentum of the particles which *leave the surface*, or in other words "were reflected," by the wall. The volume is $\mathcal{S}\dfrac{(-p_z)}{m}\Delta t$, the total momentum *taken from* the wall is equal to

$$\sum_{\text{refl}}\Delta\vec{p}_i = \frac{\mathcal{S}}{m}\Delta t\rho \int_{p_z<0} (-p_z)\vec{p}f(\vec{p})d^3\vec{p} \tag{4.8}$$

The *net total momentum* transferred to the wall is obtained by integrating on all values of momentum p_z, i.e.,

$$\sum_{\text{inc}} \Delta \vec{p_i} - \sum_{\text{refl}} \Delta \vec{p_i} = \frac{S}{m} \Delta t \rho \int p_z \vec{p} f(\vec{p}) d^3 \vec{p} \tag{4.9}$$

Because the probability is an even function, the integrals in $p_x p_z$ and in $p_y p_z$ vanish, so that the integral reduces to the z component, normal to the surface :

$$\left(\sum_{\text{inc}} \Delta \vec{p_i} - \sum_{\text{rEfl}} \Delta \vec{p_i} \right)_z = 2S \Delta t \rho \int \frac{p_z^2}{2m} f(\vec{p}) d^3 \vec{p} \tag{4.10}$$

It is related to the average value of the contribution to the kinetic energy of the motion along Oz : since the distribution of momenta is taken as isotropic, this is one third of the average value of the total kinetic energy E_c. One can deduce the expression of the normal force on the wall and thus of the pressure :

$$F_{\text{total}} = 2S\rho \frac{\langle E_c \rangle}{3} \tag{4.11}$$

$$P = 2\rho \frac{\langle E_c \rangle}{3} \tag{4.12}$$

As you already know, and will be again shown in this chapter in §4.4.2, the average kinetic energy per particle in a monoatomic ideal gas, with only translational degrees of freedom, is equal to

$$\langle E_c \rangle = \left\langle \frac{p^2}{2m} \right\rangle = \left\langle \frac{p_x^2 + p_y^2 + p_z^2}{2m} \right\rangle = \frac{3}{2} k_B T \tag{4.13}$$

Consequently, the pressure is related to the particles density and to the temperature through

$$P = 2\rho \frac{1}{3} \cdot \frac{3}{2} k_B T = \rho k_B T = \frac{N}{\Omega} k_B T \tag{4.14}$$

a relation which is the well-known ideal gas law.

This type of kinetic calculation will be used in chapter 9 on thermal radiation; in the case of nonequilibrium Statistical Physics, it also allows one to estimate the order of magnitude of transport coefficients (for example, thermal conductivity, viscosity).

4.3 Classical or Quantum Statistics ?

We now consider the range of validity of *Classical* Statistical Physics for studying the properties of a gas of free molecules.

It is well known that the correct treatment of the particles motion must be done in Quantum Mechanics, so that the motion of a particle should not be described in terms of a trajectory but rather of a probability of presence deduced from a wave function. Moreover, the specific question raised by the Quantum Statistics, as will be seen in Chapter 5, is that of the indistinguishability of the particles.

But you also learnt that Classical Mechanics can however be used when the size of the considered system, here the average distance between two particles, is much larger than the characteristic dimension of the wave function, here its wavelength (in the opposite situation, the description in Quantum Mechanics is mandatory).

Let us first evaluate the characteristic dimension, at a given temperature T, of the wavelength of the molecules' wave function. The momenta distribution is a function of T : it was just recalled that the mean kinetic energy per particle of an ideal gas is equal to $\frac{3}{2}k_BT$ in thermal equilibrium at temperature T, which gives

$$\sqrt{\langle p^2 \rangle} = \sqrt{3mk_BT} \tag{4.15}$$

Using the Louis de Broglie relation (1926),

$$\lambda = \frac{h}{p} \tag{4.16}$$

to the root-mean-square value (4.15) of the momentum, one associates the wavelength λ of a free particle wave function, where h is the Planck constant. In order to simplify the forthcoming expressions of § 4.4, the *thermal de Broglie wavelength* is defined here by

$$\lambda_{\text{th}} = \frac{\sqrt{3}}{\sqrt{2\pi}} \frac{h}{\sqrt{\langle p^2 \rangle}} = \frac{h}{\sqrt{2\pi mk_BT}} \tag{4.17}$$

To know which of the Classical or Quantum Statistics applies, one thus compares λ_{th} to the average distance d between particles, of the order of $(\Omega/N)^{1/3}$.

If $\lambda_{\text{th}} \ll d$, $\lambda_{\text{th}}^3 \ll \frac{\Omega}{N}$, the classical statistics is applicable. Then

$$\frac{h}{\sqrt{2\pi mk_BT}} \ll \left(\frac{\Omega}{N}\right)^{1/3} \tag{4.18}$$

$$h \ll \left(\frac{\Omega}{N}\right)^{1/3} \sqrt{2\pi mk_BT} \tag{4.19}$$

This is a **high temperature or low density** regime.

On the other hand, for $\lambda_{\text{th}} \geq d$, the treatment of the present chapter is no longer valid and a quantum analysis is necessary (see Chapter 5 and following ones).

Take the example of a nitrogen gas in standard conditions of temperature and pressure : $\Omega = 22.4$ liters for $\mathcal{N} = 6.02 \times 10^{23}$ molecules and $T = 273$ K. The quantity to be compared to $h = 6.63 \times 10^{-34}$ J.sec is equal here to 1×10^{-31}. Indeed the regime is classical. On the other hand, when at constant pressure the temperature is lowered to a few kelvins, the thermal L. de Broglie wavelength and the average distance between particles become comparable for gases of low atomic mass like helium (under the condition that they are not solid under these conditions!). In fact the right member of (4.19) is also written $(k_B T/P)^{1/3}(2\pi m k_B T)^{1/2}$ and thus varies like $m^{1/2}T^{5/6}$. Classical Statistics is then no longer valid (see chapter 9).

In the next section we will study the ideal gas by the **classical** "Maxwell-Boltzmann" statistics and thus assume that the condition $\lambda_{\text{th}} \ll d$ is fulfilled. Consequently, a classical description of the motion and of the phase space can be used (see § (1.1)) : until the next chapter we forget Quantum Mechanics...

4.4 Classical Statistics Treatment of the Ideal Gas in the Canonical Ensemble

As explained in chapter 2, the statistical treatment of a macroscopic system will give the same physical results, whether performed in the microcanonical, canonical or grand canonical ensemble. In the ideal gas case, the microcanonical ensemble leads to somewhat more complicated calculations. The treatment could be done in the grand canonical ensemble; here we will choose the canonical ensemble. We follow the general method (see chapter 3) consisting in first calculating the canonical partition function Z_c (§ 4.4.1), then deducing the average energy (§ 4.4.2) and the free energy F (§ 4.4.3) the partial derivatives of which provide the thermodynamical parameters pressure, entropy and chemical potential.

4.4.1 Calculation of the Canonical Partition Function

We are concerned here by N monoatomic molecules, confined into a volume Ω, with energies reduced to their translational kinetic energies within this volume. Consequently, the probability of occurrence of the state in which the molecule i is within the volume $d^3\vec{r}_i$ around $\vec{r}_i$, with the momentum $\vec{p}_i$ to

$d^3\vec{p}_i$, is given by

$$\frac{1}{Z_c} \exp\left[-\beta \sum_i \left(\frac{p_i^2}{2m} + V(\vec{r}_i)\right)\right] \prod_i \frac{C_N d^3\vec{r}_i d^3\vec{p}_i}{h^{3N}} \tag{4.20}$$

To express the confinement, the potential energy is taken as zero inside the volume Ω, infinite outside it. The canonical partition function for N particles, which normalizes the probabilities to unity, is calculated using the measure of the $6N$-dimensional phase space introduced in § 1.2.2 and expressed in the above formula. (Remember that C_N is a constant which only depends on the number of particles.)

$$Z_c = \int \exp\left(-\beta \sum_i \left(\frac{p_i^2}{2m} + V(\vec{r}_i)\right)\right) \prod_i \frac{C_N d^3\vec{r}_i d^3\vec{p}_i}{h^{3N}} \tag{4.21}$$

As already seen in § 2.4.5 of chapter 2, since the total energy consists in a sum of similar one-particle terms, the partition function is a product of terms :

$$Z_c = C_N (z_1)^N \tag{4.22}$$

with

$$z_1 = \int \exp\left[-\beta\left(\frac{\vec{p}_1^2}{2m} + V(\vec{r}_1)\right)\right] \frac{d^3\vec{r}_1 d^3\vec{p}_1}{h^3} \tag{4.23}$$

$$= \int e^{-\beta V(\vec{r}_1)} d^3\vec{r}_1 \int e^{-\beta p_1^2/2m} \frac{d^3\vec{p}_1}{h^3} \tag{4.24}$$

(Do not forget the C_N factor of the phase space measure!)

If the potential only expresses that the molecule remains within the volume Ω, the integral in $\vec{r}_1$ is equal to Ω (the exponential is equal to one in the allowed volume, zero outside it).

Note : If the potential is position-dependent, as in the case of a gas in equilibrium in the gravity field, the probability of presence and the density both vary in $\exp(-\beta V(\vec{r}))$ and the integral in $\vec{r}_1$ is different from Ω (it is proportional to the characteristic extension of the potential energy).

From now we assume that the gas is confined, with a uniform probability of presence within the allowed volume.

The integral in momentum is reduced into the product of three Gaussian integrals, each one concerning a single coordinate :

$$\int \exp\left(-\beta \frac{\vec{p}_1^2}{2m}\right) \frac{d^3\vec{p}_1}{h^3} = \left[\int_{-\infty}^{+\infty} \exp\left(-\beta \frac{p_{1x}^2}{2m}\right) \frac{dp_{1x}}{h}\right]^3 \tag{4.25}$$

Homogeneity considerations predict a result with a dimension $(\text{momentum}/h)^3$. Now the characteristic momentum in the exponential is $\sqrt{\frac{m}{\beta}} = \sqrt{mk_BT}$. The mathematical appendix at the end of this book gives the value of the dimensionless Gaussian integral. Finally

$$z_1 = \frac{\Omega}{h^3}(2\pi mk_BT)^{3/2} = \frac{\Omega}{\lambda_{\text{th}}^3} \tag{4.26}$$

where the thermal de Broglie wavelength, defined in § 4.3, is expressed. One deduces the N-particle canonical partition function

$$Z_c = C_N\left(\frac{\Omega}{h^3}\right)^N(2\pi mk_BT)^{3N/2} = C_N\left(\frac{\Omega}{\lambda_{\text{th}}^3}\right)^N \tag{4.27}$$

that we will use to calculate the free energy F. But before that we will calculate the average value of the N-molecule *kinetic* energy, using Maxwell's original method.

4.4.2 Average Energy ; Equipartition Theorem

In the absence of potential energy other than the confinement one, the *internal* energy in the sense of Thermodynamics reduces to this average kinetic energy. By definition :

$$\langle E_c\rangle = \frac{\int\sum_i\frac{\vec{p}_i^{\,2}}{2m}\exp\left[-\beta(\sum_j\frac{\vec{p}_j^{\,2}}{2m}+V(\vec{r}_j))\right]\prod_i\frac{C_Nd^3\vec{r}_id^3\vec{p}_i}{h^{3N}}}{\int\exp\left[-\beta(\sum_i\frac{\vec{p}_i^{\,2}}{2m}+V(\vec{r}_i))\right]\prod_i\frac{C_Nd^3\vec{r}_id^3\vec{p}_i}{h^{3N}}} \tag{4.28}$$

The integrals of this complicated expression factorize both in the numerator and in the denominator, the particle i playing a specific role :

$$\langle E_c\rangle = \sum_i\int\frac{d^3\vec{r}_id^3\vec{p}_i}{h^3}\frac{\vec{p}_i^{\,2}}{2m}\exp\left[-\beta(\frac{\vec{p}_i^{\,2}}{2m}+V(\vec{r}_i))\right] \tag{4.29}$$

$$\times\frac{\prod_{j\neq i}\int\frac{C_Nd^3\vec{r}_jd^3\vec{p}_j}{h^3}\exp\left[-\beta(\frac{\vec{p}_j^2}{2m}+V(\vec{r}_j))\right]}{\prod_j\int\frac{C_Nd^3\vec{r}_jd^3\vec{p}_j}{h^3}\exp\left[-\beta(\frac{\vec{p}_j^{\,2}}{2m}+V(\vec{r}_j))\right]}$$

As for particle i, one separates the terms corresponding to the three degrees of freedom x_i, y_i and z_i. A sum of $3N$ analogous terms appears :

$$\langle E_c\rangle = 3N\frac{\int_{-\infty}^{+\infty}\frac{p_x^2}{2m}\exp\left(-\beta\frac{p_x^2}{2m}\right)dp_x}{\int_{-\infty}^{+\infty}\exp\left(-\beta\frac{p_x^2}{2m}\right)dp_x} \tag{4.30}$$

For each of these terms, an integration by parts allows one to transform the numerator term into the denominator one to the factor $\dfrac{1}{2\beta} = \dfrac{k_B T}{2}$, for

$$\int_{-\infty}^{+\infty} \frac{p_x^2}{2m} \exp\left(-\beta \frac{p_x^2}{2m}\right) dp_x = \frac{1}{2\beta} \int_{-\infty}^{+\infty} \exp\left(-\beta \frac{p_x^2}{2m}\right) dp_x \qquad (4.31)$$

Consequently,

$$U = \frac{3N}{2\beta} = \frac{3}{2} N k_B T \qquad (4.32)$$

In fact this calculation is valid in the Maxwell-Boltzmann Classical Statistics for the average value of any energy term expressed as a *quadratic* degree of freedom, for example an elastic potential energy of the type $\dfrac{1}{2} k_i x_i^2$, for which $k_i > 0$. Indeed one will always perform an integration by parts of a second-degree term in energy in the numerator to express the corresponding factor of the denominator partition function, which will provide the factor $k_B T/2$. But this property *is not valid* for a polynomial of degree different than two, for example for an anharmonic term in x_i^3.

This result constitutes the "theorem of energy equipartition", established by James C. Maxwell as early as 1860 :

> *For a system in canonical equilibrium at temperature T, the average value of the energy per particle and per <u>quadratic</u> degree of freedom is equal to* $\dfrac{1}{2} k_B T$.

We have already used this property several times, in particular in §4.3 to find the root-mean-square momentum of the ideal gas at temperature T. It is also useful to determine the thermal fluctuations of an oscillation, providing a potential energy in $\dfrac{1}{2} k x^2$ for a spring, in $\dfrac{1}{2} C \vartheta^2$ for a torsion pendulum, and more generally to calculate the average value at temperature T of a hamiltonian of the form

$$h = \sum_i a_i x_i^2 + \sum_j b_j p_j^2 \qquad (4.33)$$

where a_i is independent of x_i, b_j does not depend of p_j, and all the coefficients a_i and b_j are positive (which allows to use this property for an energy development around a *stable* equilibrium state).

A case of application of the energy equipartition theorem is the specific heat at constant volume associated to U. We recall its expression

$$C_v = \left(\frac{\partial U}{\partial T}\right)_{\Omega, N} \qquad (4.34)$$

In the case of a monoatomic ideal gas, C_v only contains the contribution of the three translation degrees of freedom and is thus equal to $\frac{3}{2}Nk_B$.

4.4.3   Free Energy ; Physical Parameters (P, S, μ)

We saw in chapter 3 that the canonical partition function Z_c is related to the free energy through

$$F = U - TS = -k_B T \ln Z_c \tag{4.35}$$

In the case of an ideal gas, the differential of F is

$$dF = -Pd\Omega - SdT + \mu dN \tag{4.36}$$

The volume Ω and the number of particles N are extensive variables, the temperature T is intensive. The chemical potential, already introduced in § 3.5.3, is here the variation of free energy of the system at fixed temperature and volume when one particle is added.

Substituting the expression of Z_c obtained in § 4.4.1, one obtains :

$$F = -k_B T \left(N \ln \frac{\Omega}{h^3}(2\pi m k_B T)^{3/2} + \ln C_N \right) \tag{4.37}$$

$$F = -N k_B T \left(\ln \frac{\Omega}{Nh^3}(2\pi m k_B T)^{3/2} + \ln N + \frac{1}{N} \ln C_N \right) \tag{4.38}$$

For the potential F to be extensive, the quantity between brackets must be intensive. This dictates that the ratio $\dfrac{\Omega}{N}$ should appear in the first term and that the remaining terms should be constant :

$$\ln N + \frac{1}{N} \ln C_N = \text{constant} \tag{4.39}$$

i.e.,

$$\ln C_N = -N \ln N + \text{constant} \times N \tag{4.40}$$

From now on, we will take

$$C_N = \frac{1}{N!}, \text{ i.e. } \ln C_N = -N \ln N + N + O(\ln N) \tag{4.41}$$

so that the constant in Eq. (4.39) is equal to unity. This choice will be justified in § 4.4.4 by a physical argument, the Gibbs paradox, and will be demonstrated in § 6.7. Then

$$F = -N k_B T \left[\ln \left(\frac{\Omega}{Nh^3}(2\pi m k_B T)^{3/2} \right) +1 \right] = -N k_B T \ln \left(\frac{e\Omega}{N\lambda_{\text{th}}^3} \right) \tag{4.42}$$

One deduces the pressure, the entropy and the chemical potential of the assembly of N particles :

$$P = -\left(\frac{\partial F}{\partial \Omega}\right)_{N,T} = \frac{Nk_BT}{\Omega} \tag{4.43}$$

$$S = -\left(\frac{\partial F}{\partial T}\right)_{N,\Omega} = Nk_B \ln\left(\frac{\Omega}{N\lambda_{\text{th}}^3}\right) + \frac{5}{2}Nk_B \tag{4.44}$$

$$\mu = \left(\frac{\partial F}{\partial N}\right)_{T,\Omega} = -k_BT \ln\left(\frac{\Omega}{N\lambda_{\text{th}}^3}\right) \tag{4.45}$$

One identifies the first of these relations as the ideal gas law, that we now just re-demonstrated using Statistical Physics.

The expression obtained for $S(\Omega, N, T)$ is extensive but does not satisfy the Nernst principle, since it does not vanish when the temperature tends to zero : the entropy would become very negative if the de Broglie thermal wavelength was becoming too large with respect to the mean distance between particles. This confirms what we already expressed in § 4.3, that is, the ideal gas model is no longer valid at very low temperature.

In the validity limit of the ideal gas model, $\lambda_{\text{th}} \ll d$ is equivalent to $\Omega/N\lambda_{\text{th}}^3$ very large, thus to μ very negative ($\mu \to -\infty$).

Remember that the measure of the classical N-particle phase space must be taken equal to $\dfrac{1}{N!}\displaystyle\prod_{i=1}^{N}\dfrac{d^3\vec{r}_i d^3\vec{p}_i}{h^3}$ to ensure the extensivity of the thermodynamical potentials F and S. This will be justified in chapter 6 : the factor C_N expresses the common classical limit of both Quantum Statistics.

4.4.4   Gibbs Paradox

At the end of the 19th century, John Willard Gibbs used the following physical argument to deduce the requirement of the constant $C_N = \frac{1}{N!}$ in the expression of the infinitesimal volume of the phase space : consider two containers of volumes Ω_1 and Ω_2, containing under the same density and at the same temperature T, respectively, N_1 and N_2 molecules of gas (Fig. 4.3). One then opens the partition which separates the two containers, the gases mix and the final total entropy is compared to the initial total one. If the gases in the two containers were of different chemical natures, the disorder and thus the entropy increase. On the other hand, if both gases are of the same nature and the molecules indistinguishable, the entropy is unchanged in the process.

The detailed calculation of the different terms in the total entropy, *before*

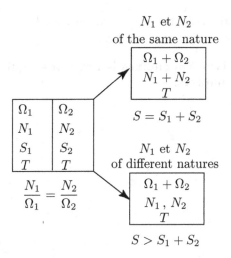

FIG. 4.3: Mixture of two gases at the same temperature and with the same density.

and *after* opening the partition, is simple and uses the partial derivative with respect of temperature of expression (4.38) of the free energy, including the constant C_N : if one assumes that $C_N = 1$, there will be a variation of the total entropy even for a mixture of molecules of the same nature. When one introduces the factor $C_N = 1/N!$, the result corresponds to the physical prediction. The choice of this factor is not surprising once one considers that molecules of the same chemical nature are *indistinguishable* : for a system of N indistinguishable particles, there are $N!$ possible permutations of these particles among their one-particle states, providing the same macroscopic state. In order to account only once for this state in the enumeration that will lead to the physical predictions, among others of the entropy, it is natural to choose the value $C_N = 1/N!$.

4.5   Conclusion

Using Statistical Physics, we have been able to find again the well-known results of Thermodynamics on the ideal gas and to justify them. It seems interesting, now that we come to the end of the study of Classical Statistics, to present the respective contributions of Thermodynamics (Appendix 4.1) and of Statistical Physics (Appendix 4.2) on a practical example. This will be done on the analysis of the chemical reactions.

In the Statistical Physics part, one will show in details the various microscopic terms of the hamiltonian of a gas of diatomic molecules; this allows one to understand the origin of a law very commonly used in Chemistry, that is, the law of mass action. Enjoy your reading!

Reading these two Appendixes is not a prerequisite to understanding the next chapters.

Summary of Chapter 4

First we have presented some orders of magnitude, relative to nitrogen molecules in standard conditions of temperature and pressure. They justify that the hypotheses of the ideal gas, of molecules with no mutual interaction, is reasonable : indeed the distances traveled by a molecule between collisions are very large with respect to a molecule radius and the characteristic time to reach equilibrium is very short as compared to a typical time of measurement.

Then the kinetic calculation of a gas pressure was recalled.

The domain of validity of the ideal gas model, in the framework of both Classical Mechanics and Classical Statistics, is the range of high temperatures and low densities : the criterion to be fulfilled is

$$h \ll \left(\frac{\Omega}{N}\right)^{1/3} \sqrt{2\pi m k_B T}, \text{ or } \lambda_{\text{th}} = \frac{h}{\sqrt{2\pi m k_B T}} \ll d = \left(\frac{\Omega}{N}\right)^{1/3}$$

where λ_{th} is the thermal de Broglie wavelength and d the mean distance between particles.

The statistical treatment of the ideal gas is done in this chapter in the canonical ensemble, the N-particle canonical partition function is given by

$$Z_c = C_N \left[\left(\frac{\Omega}{h^3}\right)(2\pi m k_B T)^{3/2}\right]^N$$

To ensure that the free energy and the entropy are extensive and to solve the Gibbs paradox, one must take $C_N = 1/N!$.

The average value at temperature T of the energy associated to a *quadratic* degree of freedom is equal to $\frac{1}{2}k_B T$ (theorem of the equipartition of energy). This applies in particular to the kinetic energy terms for each coordinate.

The chemical potential of the ideal gas is very negative; its entropy does not satisfy to the third law of Thermodynamics (but the ideal gas model is no longer valid in the limit of very low temperatures).

Appendix 4.1

Thermodynamics and Chemical Reactions

We will consider here macroscopic systems composed of several types of molecules or chemical species, possibly in several phases : one can study a chemical reaction between gases, or between gases and solids, etc. The first question that will be raised is "What is the spontaneous evolution of such systems ?" ; then the conditions of chemical equilibrium will be expressed. These types of analyses, only based on Thermodynamics, were already introduced in your previous studies. Now Appendix 4.2 will evidence the contribution of Statistical Physics.

1. Spontaneous Evolution

In the case of a thermally *isolated* system, the second law of Thermodynamics indicates that a spontaneous evolution will take place in a direction corresponding to an increase in entropy S, S being maximum in equilibrium.

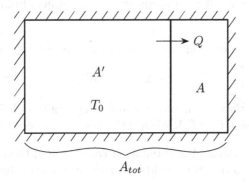

FIG. 4.4: The system A is in thermal contact with the heat reservoir A', with which it exchanges the heat Q. The combined system is isolated.

Chemical reactions are mostly done at *constant temperature* : one then has to consider the case of a system A in thermal contact with a heat reservoir A' at temperature T_0 (Fig. 4.4). The combined system A_{tot}, made of A and A' is an isolated system, the entropy S_{tot} of which increases in a spontaneous process :

$$\Delta S_{tot} \geq 0 \tag{4.46}$$

The entropy variation of the combined system

$$\Delta S_{tot} = \Delta S + \Delta S' \tag{4.47}$$

can be expressed as a function of parameters of the system A alone.

Indeed, if the amount of heat Q is given to A by the heat reservoir, the reservoir absorbs $-Q$ and its entropy variation is equal to $\Delta S' = -Q/T_0$: the transfer of Q does not produce a variation of T_0 and one can imagine a reversible sequence of isothermal transformations at T_0 producing this heat exchange.

Besides, the internal energy variation in A, i.e., ΔU, is, from the first law of Thermodynamics,

$$\Delta U = W + Q \tag{4.48}$$

where W is the work operated in the considered transformation. Consequently,

$$\Delta S_{tot} = \Delta S - \frac{(\Delta U - W)}{T_0} = \frac{-\Delta \mathcal{F}_0 + W}{T_0} \tag{4.49}$$

where, by definition, $\mathcal{F}_0 = U - T_0 S$, which is identical to the Helmholtz free energy $F = U - TS$ of system A if the temperature of A is that of the heat reservoir (in an off-equilibrium situation T differs from T_0 and $\mathcal{F}$ is not identical to F).

From the condition of spontaneous evolution one deduces that

$$-\Delta \mathcal{F}_0 \geq (-W) \tag{4.50}$$

which expresses that the maximum work that may be delivered by A in a reversible process is $(-\Delta \mathcal{F}_0)$, whence the name of "free energy."

If the volume of system A is maintained constant and there is no other type of work, $W = 0$, so that the condition for spontaneous evolution is then given by

$$\Delta \mathcal{F}_0 \leq 0 \tag{4.51}$$

This condition for a system in contact with a heat reservoir replaces the condition $\Delta S_{\text{tot}} \geq 0$ for an isolated system.

A more realistic situation is the one in which system A is kept at *constant temperature* and *constant pressure*. This is the case in laboratory experiments when a chemical reaction is performed at fixed temperature T_0 and under the atmospheric pressure P_0 : system A is exchanging heat and work with a very large reservoir A', which can thus give energy to A without any modification of T_0 and also exchange a volume $\Delta\Omega$ with A without any change of its pressure P_0 (Fig. 4.5).

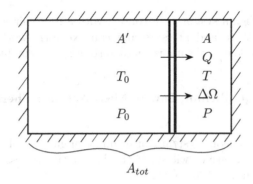

FIG. 4.5: System A is in contact with the larger system A', with which it is exchanging the amount of heat Q and the volume $\Delta\Omega$.

Similarly, one considers that the combined system made of A and A' is isolated, so that in a spontaneous process

$$\Delta S_{\text{tot}} = \Delta S + \Delta S' \geq 0 \tag{4.52}$$

If A absorbs Q from A', $\Delta S' = -Q/T_0$, where T_0 is again the temperature of A'.

The variation of internal energy of A includes the received heat Q, the mechanical work $-P_0\Delta\Omega$ received by A from A', and also W^*, work of any other origin (electrical for example) that A may have received :

$$\Delta U = Q - P_0\Delta\Omega + W^* \tag{4.53}$$

Consequently,

$$\Delta S_{\text{tot}} = \Delta S - \frac{Q}{T_0} = \frac{T_0\Delta S - (\Delta U + P_0\Delta\Omega - W^*)}{T_0} \tag{4.54}$$

i.e., $$\Delta S_{\text{tot}} = \frac{-\Delta\mathcal{G}_0 + W^*}{T_0} \tag{4.55}$$

Here we defined $\mathcal{G}_0 = U + P_0\Omega - T_0S$, which identifies to the Gibbs free enthalpy of system A, that is, $G = U + P\Omega - TS$, when $P = P_0$ and $T = T_0$.

A spontaneous evolution will be expressed through the condition

$$-\Delta\mathcal{G}_0 + W^* \geq 0 \tag{4.56}$$

When all the external parameters of system A, except its volume, are kept fixed, W^* is null and the condition of spontaneous evolution is then

$$\Delta\mathcal{G}_0 \leq 0 \tag{4.57}$$

Consequently, when a system is in contact with a reservoir so that it remains at fixed temperature and pressure and can exchange work only with this reservoir, the stable equilibrium is characterized by the condition of minimum of $\mathcal{G}_0$.

2. Chemical Equilibrium, Law of Mass Action : Thermodynamical Approach

Consider a homogeneous system (for example in gaseous phase) with several types of molecules : here we will study the dissociation reaction of iodine vapor at temperatures in the 1000 K range :

$$I_2 \rightleftarrows 2I$$

In this reaction the number of iodine atoms is conserved between the species corresponding to the two members of the chemical equation, which can also be expressed in the form

$$2I - I_2 = 0$$

the species disappearing in the reaction carrying a negative sign, the one appearing a positive one. In the same way, the most general chemical equilibrium will be written

$$\sum_{i=1}^{n} b_i N_i = 0 \tag{4.58}$$

The b_i are the integral coefficients of the balanced reaction for the B_i molecules. The number N_i of *molecules* of B_i type in the system is changed if the equilibrium is displaced but the total number of *atoms* of each species in the system is conserved. The variations of the numbers N_i are thus proportional to the coefficients b_i of the chemical equilibrium

$$dN_i = b_i\, d\lambda \quad \text{for any } i$$

where $d\lambda$ is a proportionality constant. dN_i is positive for the molecules appearing, negative for those vanishing in the reaction.

Consider a chemical equilibrium at fixed temperature T and pressure P, thus characterized by $dG(T, P) = 0$. In such conditions the equilibrium is expressed by :

$$dG = \sum_{i=1}^{n} dG_i = \sum_{i=1}^{n}(-S_i dT + \Omega_i dP + \mu_i dN_i)$$

$$\text{At fixed } T \text{ and } P, \ dG = \sum_{i=1}^{n} \mu_i dN_i = 0 \tag{4.59}$$

Here $\mu_i = \left(\dfrac{\partial G}{\partial N_i}\right)_{T,P}$ is the chemical potential of species B_i. Replacing dN_i by its expression versus λ, one obtains the relation characterizing the equilibrium :

$$\sum_{i=1}^{n} b_i \mu_i = 0 \tag{4.60}$$

The chemical potentials depend on the experimental conditions, in particular on the concentrations of the different species. Assume that these reagents are *ideal gases* and look for the dependence of μ_i on the partial pressure P_i of one of these gases.

Since G_i is extensive and depends on a single extensive parameter N_i, the number of molecules of the species i, then one must have

$$G_i = \mu_i N_i \tag{4.61}$$

The dependence in temperature and pressure of μ_i is thus that of G_i. For an ideal gas, at fixed temperature,

$$G_i = U_i + P_i \Omega - T S_i \tag{4.62}$$

introducing the partial pressure P_i of species i, such that $P_i \Omega = N_i k_B T$. Consequently,

$$dG_i = -S_i dT + \Omega dP_i + \mu_i dN_i \tag{4.63}$$

$$\left(\frac{\partial G_i}{\partial P_i}\right)_{T,N_i} = N_i \left(\frac{\partial \mu_i}{\partial P_i}\right)_{T,N_i} = \Omega = \frac{N_i k_B T}{P_i} \tag{4.64}$$

i.e., by integrating (4.64),

$$\mu_i = \mu_i^0(T) + k_B T \ln \frac{P_i}{P_0} \tag{4.65}$$

where $\mu_i^0(T)$ is the value of the chemical potential of species i at temperature T and under the standard pressure P_0.

The general condition of chemical equilibrium is thus given by

$$\sum_{i=1}^{n} b_i \mu_i^0 + k_B T \sum_{i=1}^{n} \ln \left(\frac{P_i}{P_0} \right)^{b_i} = 0 \qquad (4.66)$$

One writes

$$\frac{\Delta G_m^0}{\mathcal{N}} = \sum_{i=1}^{n} b_i \mu_i^0 \qquad (4.67)$$

ΔG_m^0, the molar free enthalpy at the temperature T of the reaction, is the difference between the molar free enthalpy of the synthesized products and that of the used reagents, $\mathcal{N}$ is the Avogadro number.

One also defines

$$K_p = \prod_{i=1}^{n} \left(\frac{P_i}{P_0} \right)^{b_i} \qquad (4.68)$$

For example, in the dissociation equilibrium of iodine molecules,

$$K_p = \frac{(P_I/P_0)^2}{(P_{I_2})/P_0} \qquad (4.69)$$

One recognizes in K_p the constant of mass action.

The above equation gives

$$K_p = e^{-\Delta G_m^0 / \mathcal{R}T} \qquad (4.70)$$

where $\mathcal{R} = \mathcal{N}k_B$ is the ideal gases constant, $\mathcal{R} = 8.31$ J.K^{-1}. From the tables of free enthalpy values *deduced from experiments*, one can deduce the value of K_p. From the experimental study of a chemical equilibrium at temperature T, one can also deduce the value of K_p, and thus of ΔG_m^0, at temperature T : if the temperature of the experiment is modified from T to T', the partial pressures are changed, $K_p(T)$ becomes $K_p(T')$, such that

$$\ln K_p(T') = -\Delta G_m^0(T')/\mathcal{R}T' \qquad (4.71)$$

By definition, at a given T, $\Delta G_m^0 = \Delta H^0 - T\Delta S^0$, where ΔH^0 is the molar enthalpy of the reaction and ΔS^0 its entropy variation. If T and T' are close enough, one makes the approximation that neither ΔH^0 nor ΔS^0 noticeably vary between these temperatures. Then

$$\ln K_p(T') - \ln K_p(T) = -\Delta H^0 \left(\frac{1}{\mathcal{R}T'} - \frac{1}{\mathcal{R}T} \right) \qquad (4.72)$$

i.e., also

$$\frac{d\ln K_p(T)}{dT} = \frac{\Delta H^0}{\mathcal{R}T^2} \tag{4.73}$$

From the enthalpy ΔH^0, one can predict the direction of displacement of the equilibrium when the temperature increases : if $\Delta H^0 > 0$, that is, the enthalpy increases in the reaction and the reaction is endothermal, the equilibrium is thus displaced toward the right side when T increases. This is in agreement with the principle of Le Châtelier :

> If a system is in stable equilibrium, any spontaneous change of its parameters must lead to processes which tend to restore the equilibrium.

As an application, the experimental data below on the dissociation of iodine allow to deduce the enthalpy of this reaction in the vicinity of 1000 K.

For a fixed total volume of 750.0 cm^3, in which a total number n of molecules I_2 was introduced at temperature T, the equilibrium pressure is P :

T (K)	973	1 073	1 173
$P(10^{-2}$ atm)	6.24	7.50	9.18
$n(10^{-4}$ moles)	5.41	5.38	5.33

As an exercise, one can calculate K_p at each temperature and verify that $\Delta H^0 = 157$ kJ.mole^{-1}. This reaction is endothermal.

The example presented here suggests the rich domain, but also the limitations of the applications of Thermodynamics in Chemistry : Thermodynamics does not reduce to a game between partial derivatives, as some students may believe! It allows one to predict the evolution of new reactions, from *tables deduced from measures*. However, only the contribution of Statistical Physics allows us to predict these evolutions from the sole spectroscopic data and *microscopic models*. It is this latter approach that will be followed now. Thanks to the understanding acquired in this first part of the Statistical Physics course, we will be able to deduce the constant of mass action from the quantum properties of the chemical species of the reaction.

Appendix 4.2

Statistical Physics and Chemical Reactions

The same example of dissociation of the iodine molecule at 1000 K, in gas phase, will now be treated by applying the general methods of Statistical Physics. The calculations will be completely developed, which may be regarded as tedious but yet they are instructive!

1. Energy States : The Quantum Mechanics Problem

Iodine I is a monoatomic gas, its only degrees of freedom are those of *translation*. The *electronic* structure of its fundamental state is $^2P_{3/2}$ (notation $^{2S+1}L_J$: orbital state $\ell = 1$ whence the notation P; spin $1/2$, i.e., $2S+1 = 2$; total kinetic moment equal to $J = 3/2$, therefore the degeneracy 4). Iodine has a nuclear moment J=5/2.

The hamiltonian of the iodine atom in the gas phase is therefore given by

$$\mathcal{H}_I = \mathcal{H}_{I\ \text{transl}} + \mathcal{H}_{I\ \text{elec}} + \mathcal{H}_{I\ \text{nucl}} \tag{4.74}$$

The molecule I_2 has six degrees of freedom, which can be decomposed into three degrees of *translation* of its center of mass, two of *rotation* and one of *vibration* :

$$\mathcal{H}_{I_2} = \mathcal{H}_{I_2\text{transl}} + \mathcal{H}_{\text{rot}} + \mathcal{H}_{\text{vibr}} + \mathcal{H}_{I_2\text{elec}} + \mathcal{H}_{I_2\text{nucl}} \tag{4.75}$$

Since this molecule is symmetrical, one should account for the fact that the two iodine atoms are indistinguishable inside the molecule and thus the physics of the system is invariant in a rotation of 180 degrees around an axis perpendicular to the bond in its middle.

The *fundamental electronic state* of molecular iodine is not degenerate. It lies lower than the fundamental electronic state of atomic iodine. The dissociation energy, necessary at zero temperature to transform a molecule into two iodine atoms, is equal to $E_d = 1.542$ eV, which corresponds to 17,900 K : one should bring $E_d/2$ per atom to move from the state I_2 to the state I.

The *nuclear spin* of iodine has the value $\mathcal{J} = 5/2$, both under the atomic and molecular forms; one assumes that no magnetic field is applied.

The splittings between energy levels of I or I_2 are known from spectroscopic measures : Raman or infrared techniques give access to the distances between rotation or vibration levels; the distance between electronic levels are generally in the visible (from 1.5 eV to 3 eV) or the ultraviolet range.

One now expresses the eigenvalues of the different terms of $\mathcal{H}_I$ and $\mathcal{H}_{I_2}$.

a. Translation States

One has to find the hamiltonian eigenstates : in the case of the iodine atom of mass m_I,

$$\mathcal{H}_{tr} = \frac{\hat{p}^2}{2m_I} + V(\vec{r}) \tag{4.76}$$

where $V(\vec{r})$ is the confinement potential limiting the particle presence to a rectangular box of volume $\Omega = L_x L_y L_z$.

The eigenstates are plane waves $\psi = (\exp i\vec{k} \cdot \vec{r})/\sqrt{\Omega}$; the $\vec{k}$ values are quantized and for the Born-Von Kármán periodic limit conditions $[\psi(x+L_x) = \psi(x)]$ (see § 6.4.2) the $\vec{k}$ components satisfy :

$$k_x L_x = n_x 2\pi \,; \ k_y L_y = n_y 2\pi \,; \ k_z L_z = n_z 2\pi$$

with n_x, n_y, n_z positive, negative or null integers.

The corresponding energies are :

$$
\begin{aligned}
\varepsilon_{I\ tr} &= \frac{\hbar^2}{2m_I}\vec{k}^2 \\
&= \frac{h^2}{2m_I}\left[\left(\frac{n_x}{L_x}\right)^2 + \left(\frac{n_y}{L_y}\right)^2 + \left(\frac{n_z}{L_z}\right)^2\right]
\end{aligned}
\tag{4.77}
$$

The translation energies $\varepsilon_{I_2\ tr}$ of the molecule I_2 are obtained by replacing m_I by m_{I_2} in the above expression of $\varepsilon_{Itransl}$.

All these states are extremely close : if Ω is of the order of 1 cm^3 and the dimensions of the order of 1 cm, the energy of the state $n_x = 1, n_y = n_z = 0$ is given by, for I $(m_I = 127$ g$)$

$$13.6 \text{ eV} \times \frac{m_H}{m_I} \cdot \left(\frac{a_0}{L_x}\right)^2 = 13.6 \times \frac{1}{127} \times (0.53 \times 10^{-8})^2 = 3 \times 10^{-18}\text{eV !}$$

The distances between energy levels are in the same range, thus very small with respect to $k_B T$, for $T = 1000$K.

This discussion implies that the ideal gas model is valid at room temperature for I and I_2 (still more valid than for N_2, since these molecules are heavier). The partition functions for their translation states will take the form (4.26)-(4.27).

b. Rotation States

The iodine molecule is linear (Fig. 4.6), its inertia momentum around an axis normal to the bond is $I = 2m \left(\dfrac{R_e}{2} \right)^2 = \mu R_e^2$ with $\mu = m_I/2$ and R_e the equilibrium distance between the two nuclei. Since $m_I = (127/\mathcal{N})$g this momentum is particularly large at the atomic scale.

The rotation hamiltonian is

$$\mathcal{H}_{\text{rot}} = \frac{1}{2I} \hat{\vec{J}}^2 \qquad (4.78)$$

and shows that the problem is isotropic in the center of mass frame.

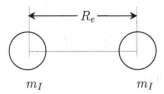

$$\text{FIG. 4.6 : Structure of the } I_2 \text{ molecule.}$$

Its eigenvalues are :

$$\varepsilon_{\text{rot}}(J) = \frac{\hbar^2 J(J+1)}{2I} = hcBJ(J+1), \quad \text{where} \quad B = \frac{\hbar}{4\pi cI} \quad \text{is given in cm}^{-1} \qquad (4.79)$$

(one recalls the definition $h\nu = hc/\lambda = hc\sigma$; the unit cm^{-1} for σ is particularly adapted to infrared spectroscopy).

The selection rules $\Delta J = \pm 1$ for allowed electric dipolar transitions between rotation levels correspond to $\varepsilon_{\text{rot}}(J+1) - \varepsilon_{\text{rot}}(J) = hc \cdot 2B(J+1)$.

For iodine $B = 0.0373$ cm^{-1}, $2hcB/k_B = 0.107$ K, to be compared to HCℓ where $B = 10.6$ cm^{-1}, $2hcB/k_B = 30.5$ K.

The rotation levels are thus still very close at the thermal energy scale.

c. Vibration States

In a first approximation the vibration hamiltonian of the molecule I_2 can be described as a harmonic oscillator hamiltonian, with a parabolic potential well centered at the equilibrium distance R_e between the two nuclei, that is, $V(R) = \frac{1}{2}K(R-R_e)^2$ (Fig. 4.7). (In this figure a realistic potential is sketched in full line; the dotted line represent its approximated harmonic potential.)

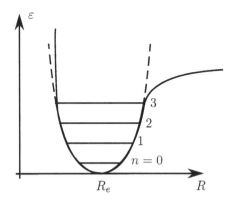

FIG. 4.7 : Potential well associated to the vibration of the molecule I_2.

$$\mathcal{H}_{\text{vibr}} = \frac{\hat{p}_R^2}{2\mu} + \frac{1}{2}K(R - R_e)^2 \tag{4.80}$$

with $\hat{p}_R = -i\hbar\dfrac{\partial}{\partial R}$, μ being the reduced mass of the system. In this approximation, the vibration energies are given by

$$\varepsilon_{\text{vibr}}(n) = \left(n + \frac{1}{2}\right)\hbar\omega \quad \text{with} \quad \omega = \sqrt{\frac{K}{\mu}} \tag{4.81}$$

The transitions $\Delta\varepsilon = \hbar\omega = hc\sigma_v$ are in the infrared domain. For I_2 $\sigma_v = 214.36$ cm^{-1}, which corresponds to $\lambda = 46.7\mu$m or $hc\sigma_v/k_B = 308$ K (for F_2 the corresponding temperature is 1280 K).

d. Electronic States

The splitting between fundamental and excited states, for both I and I_2, is of the order of a few eV, corresponding to more than 10,000 K.

The origin of the electronic energies (Fig. 4.8) is taken at the I_2 fundamental state, so that the energy of the I fundamental state is E_I, with $2E_I = E_d = 1.542$ eV, where E_d is the dissociation energy.

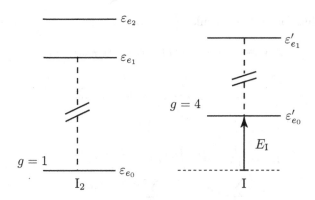

FIG. 4.8 : Electronic states of I_2 and I.

e. Iodine Nuclear Spin

It is $\mathcal{J} = 5/2$ on both species I and I_2.

2. Description of the Statistical State of the System : Calculation of the Partition Functions

It was stated in Ch.2 that for a macroscopic system the predictions for the experimental measurements of physical parameters were the same whatever the chosen statistical ensemble. Here we will work in the canonical ensemble, where the calculations are simpler.

At temperature T, the number of iodine molecules is N_{I_2}, the number of atoms produced by the dissociation of the molecules N_I. The partition function for a single iodine molecule is written z_{I_2}, that of an isolated atom z_I. Because of the particles indistinguishability, the total canonical partition function Z_c is given by :

$$Z_c = \frac{1}{N_{I_2}!}(z_{I_2})^{N_{I_2}} \frac{1}{N_I!}(z_I)^{N_I} \tag{4.82}$$

It is related to the free energy F of the system

$$F = -k_B T \ln Z_c \tag{4.83}$$

the partial derivatives of which, with respect to the numbers of particles, are the chemical potentials :

$$\mu_{I_2} = \frac{\partial F}{\partial N_{I_2}} = -k_B T \ln\left(\frac{z_{I_2}}{N_{I_2}}\right) \quad \mu_I = -k_B T \ln\left(\frac{z_I}{N_I}\right) \tag{4.84}$$

Since the total energy of an iodine atom or molecule is a sum of contributions, the partition function of a corresponding atom or a molecule is a product of factors : thus for an I_2 molecule

$$\varepsilon_{I_2} = \varepsilon_{I_2 \, \text{tr}} + \varepsilon_{\text{rot}}(J) + \varepsilon_{\text{vibr}}(n) + \varepsilon_{\text{el}} + \varepsilon_{\text{nucl}} \tag{4.85}$$

$$z_{I_2} = z_{I_2 \text{tr}} \cdot z_{\text{rot}} \cdot z_{\text{vibr}} \cdot z_{\text{el}} \cdot z_{\text{nucl}} \tag{4.86}$$

with :

$$z_{I_2 \, \text{tr}} = \frac{1}{h^3} \int d^3\vec{r} \int d^3\vec{p} \, e^{-\beta p^2 / 2m_{I_2}} \tag{4.87}$$

$$z_{\text{rot}} = \sum_J g_J e^{-\beta \varepsilon_J} = \sum_J (2J+1) e^{-\beta hc BJ(J+1)} \tag{4.88}$$

$$z_{\text{vibr}} = \sum_v g_v e^{-\beta \varepsilon_v} = \sum_n e^{-\beta \hbar \omega \left(n + \frac{1}{2}\right)} \tag{4.89}$$

$$z_{\text{el}} = \sum_e g_e e^{-\beta \varepsilon_e} \tag{4.90}$$

$$z_{\text{nucl}} = (2\mathcal{J}+1)^2 \text{ for } I_2, \quad z_{\text{nucl}} = 2\mathcal{J}+1 \text{ for } I \tag{4.91}$$

The factors g_r, g_v, g_e express the degeneracies of the rotation, vibration and electronic energy levels. The value of the nuclear spin is $\mathcal{J}$, there are two nuclei in I_2.

One will first evaluate the various partition functions, noting that a large value of the partition function is related to a large number of accessible states. The examined chemical potentials will be sums of terms, associated with the different z factors. Note that one has to calculate $\dfrac{z_{I_2}}{N_{I_2}}$ (or $\dfrac{z_I}{N_I}$) (see (4.84)), thus to divide a single factor of z_{I_2} by N_{I_2} : we will associate the factor N_{I_2} to the sole translation factor, then it no longer comes into the partition functions for the other degrees of freedom. (The indiscernability factor is thus included in the translation degree of freedom.) Then :

$$\ln \frac{z_{I_2}}{N_{I_2}} = \ln \frac{z_{\text{tr} \, I_2}}{N_{I_2}} + \ln z_{\text{rot}} + \ln z_{\text{vibr}} + \ln z_{\text{el}} + \ln z_{\text{nucl}} \tag{4.92}$$

a. Translation

The calculation of $z_{\text{tr I}}$ was already done for the ideal gas :

$$z_{\text{tr I}} = \Omega \cdot \left(\frac{2\pi m_I k_B T}{h^2} \right)^{3/2} = \frac{\Omega}{(\lambda_{\text{th I}})^3} = \frac{N_I k_B T}{P_i} \cdot \frac{1}{(\lambda_{\text{th I}})^3} \tag{4.93}$$

P_i is the partial pressure, such that $P_i \Omega = N_I k_B T$ and $\lambda_{\text{th I}}$ is the thermal de Broglie wavelength at T for I. Since we are in the classical framework, $z_{\text{tr I}}$

must be very large. Numerically λ_{th}, expressed in picometers (10^{-12} m), is given by

$$\lambda_{\text{th}} = \frac{1749}{(T_K \times m_{\text{g·mole}^{-1}})^{1/2}} \cdot 10^{-12} \text{ m}$$

At 1000 K, for the iodine atom $m_I = 127g$, $\lambda_{\text{th I}} = 4,91 \; 10^{-12}$ m.

In the same condition, for the I_2 molecule, $\lambda_{\text{th I}_2} = 3.47 \; 10^{-12}$ m.

For a partial pressure equal to the atmospheric pressure and a temperature $T = 1000$ K, one obtains

$$z_{\text{I tr}} = 1.15 \times 10^9 N_I$$

$$z_{\text{I}_2 \text{ tr}} = 3.26 \times 10^9 N_{\text{I}_2}$$

These numbers are huge!

b. Rotation

The factor $2J + 1$ is the degeneracy of the energy level $hcBJ(J + 1)$. Besides, the linear molecule I_2 is symmetrical, so that after a 180 degree rotation one gets a new state, indistinguishable from the initial one (this would not be the case for $HC\ell$ for example). One thus has to divide the above expression (4.88) by two. This is equivalent to saying that the wave function symmetry allows to keep only half of the quantum states.

Since the energy distance between rotation levels is small with respect to $k_B T$, one can replace the discrete sum by an integral

$$\frac{1}{2} \sum_J (2J + 1) e^{-\beta hcBJ(J+1)} \simeq \frac{1}{2} \int_0^\infty (2J + 1) e^{-\beta hcBJ(J+1)} dJ$$

$$\simeq \frac{1}{2} \frac{1}{\beta hcB} = \frac{k_B T}{2hcB} \tag{4.94}$$

This approximation implies that z_{rot} is a large number. Numerically, the contribution of z_{rot}, equal to $1/2\beta hcB$, is the ratio 1000 K/0.107 K=9346.

c. Vibration

$$z_{\text{vibr}} = e^{-\beta\hbar\omega/2} \sum_{n=0}^{\infty} e^{-\beta\hbar\omega n} = \frac{e^{-\beta\hbar\omega/2}}{1 - e^{-\beta\hbar\omega}} \tag{4.95}$$

At 1000 K, for I_2, the term $\beta\hbar\omega$ is 0.308 : the quantum regime is valid and this expression cannot be approximated. It has the value 3.23.

d. Electronic Term

$$z_e = \sum_e g_e e^{-\beta \varepsilon_e} \tag{4.96}$$

Owing to the large distances between electronic levels, one will only consider the fundamental level, four times degenerated for I, nondegenerated for I_2.

At 1000 K, the term z_{e_I} is equal to $4\, e^{-(E_d/2k_B T)} = 4e^{-8.94} = 5.25 \times 10^{-4}$, whereas for the same choice of energy origin $z_{e_{I_2}} = 1$.

3. Prediction of the Action of Mass Constant

Let us return to the definition relations (4.68) to (4.70) of Appendix 4.1,

$$K_p = \exp(-\Delta G_m^0/RT) = \exp(-\sum_{i=1}^{n} b_i \mu_i^0/k_B T) \tag{4.97}$$

where the μ_i^0 are the chemical potentials under the atmospheric pressure. For our example :

$$K_p = \exp[-(2\mu_I^0 - \mu_{I_2}^0)/k_B T] \tag{4.98}$$

The chemical potentials are expressed versus the partition functions calculated above :

$$\frac{1}{k_B T}(2\mu_I^0 - \mu_{I_2}^0) = -2[\ln\left(\frac{z^0 I \text{ tr}}{N_I}\right) + \ln z_{e_I}] + \ln\left(\frac{z^0 I_2 \text{ tr}}{N_{I_2}}\right) + \ln z_{\text{rot}} + \ln z_{\text{vib}} \tag{4.99}$$

Here the translation partition functions are calculated under the atmospheric pressure (standard pressure) ; the terms associated to the nuclear spins $2\ln(2\mathcal{J}+1)$ exactly compensate and $\ln z_{e\,I_2} = 0$.

From the data of the previous section one obtains :

$$\ln \frac{z^0 I \text{ tr}}{N_I} = 20.87 \quad , \quad \ln z_{eI} = -7.56$$

$$\ln \frac{z^0 I_2 \text{ tr}}{N_{I_2}} = 21.91 \quad , \quad \ln z_{\text{rot}} = 9.14 \, , \ln z_{\text{vibr}} = 1.17$$

$$K_p = e^{-5.6} \simeq 3.7 \times 10^{-3}$$

One can also predict, from the calculation of the temperature variation of K_p, a value for the enthalpy of the iodine dissociation reaction :

$$\frac{d \ln K_p}{dT} = \frac{\Delta H}{\mathcal{R} T^2} \tag{4.100}$$

i.e.,

$$\frac{d}{dT}\left(\frac{-2\mu_I^0 + \mu_{I_2}^0}{k_B T}\right) = \frac{\Delta H}{\mathcal{R} T^2} \tag{4.101}$$

ΔH is thus deduced from the logarithmic derivatives versus T, at constant P, of the different partition functions : the translation partition functions vary like $T^{5/2}$, z_{rot} like T, z_{vibr} like $\left(\sinh \frac{\beta \hbar \omega}{2}\right)^{-1}$. Consequently,

$$\frac{\Delta H}{\mathcal{R} T^2} = 2\frac{d}{dT}\left[\ln\left(\frac{z^0 I \text{ tr}}{N_I}\right) + \ln z_{e_I}\right] - \frac{d}{dT}\left[\ln\left(\frac{z^0 I_2 \text{ tr}}{N_{I_2}}\right) + \ln z_{\text{rot}} + \ln z_{\text{vibr}}\right] \tag{4.102}$$

$$= 2\left[\frac{5}{2T} + \frac{E_d}{2k_B T^2}\right] - \left[\frac{5}{2T} + \frac{1}{T} + \frac{\hbar\omega}{2k_B T^2}\coth\frac{\beta\hbar\omega}{2}\right] \tag{4.103}$$

that is,

$$\Delta H = \frac{3\mathcal{R} T}{2} + \mathcal{N}\left(E_d - \frac{\hbar\omega}{2}\coth\frac{\beta\hbar\omega}{2}\right) \tag{4.104}$$

The obtained value of ΔH is close to that of the dissociation energy per mole : $\Delta H = 152$ kJ/mole. The K_p and ΔH values, calculated from spectroscopic data, are very close to those deduced from the experiment quoted in the Appendix 4.1 ($K_p \simeq 3.4 \cdot 10^{-3}$ at 1000 K, $\Delta H = 158$ kJ/mole).

The discrepancies come in particular from the approximations done in the decomposition of the degrees of freedom of the iodine molecule : we assumed that the vibration and rotation were independent (rigid rotator model); moreover, the potential well describing vibration is certainly not harmonic.

Obviously, for an experiment performed at constant temperature T and volume Ω, the equilibrium constant K_v could have been calculated in a similar way and the free energy F would have been used. The variation of K_v with temperature provides the variation ΔU of internal energy in the chemical reaction.

4. Conclusion

Through these examples, which were developed in detail, the contribution of Statistical Physics in the prediction of chemical reactions was shown. As just seen now, if the Statistical Physics concepts refer to very fundamental notions, this discipline (sometimes at the price of lengthy calculations!) brings useful data to solve practical problems.

Chapter 5

Indistinguishability, the Pauli Principle (Quantum Mechanics)

The preceding chapter presented properties of the ideal gas which are valid in the limit where the average value, at temperature T, of the de Broglie wavelength is small with respect to the average distance between the particles in motion. Another way to express the same concept is to consider that the wave functions of two particles have a very small probability of overlapping in such conditions.

From now on, we will work in the other limit, in which collisions between particles of the same nature are very likely. In the present chapter the *Quantum Mechanics* properties applicable to this limit will be recalled. In Quantum Mechanics courses (see for example the one by J.-L. Basdevant and J. Dalibard) it is stated that, for indistinguishable particles, the Pauli principle must be added to the general postulates of Quantum Mechanics. In fact here we are only re-visiting this principle but ideas are better understood when they are tackled several times! In chapter 6 the consequences in *Statistical Physics* will be drawn on the properties obtained in the present chapter 5.

In the introduction (§ 5.1) we will list very different physical phenomena, associated with the property of indistinguishability, which will be interpreted at the end of the chapter in the framework of Statistical Physics. To introduce the issue, in § 5.2 some properties of the quantum states of two, then several, identical particles, will be recalled. In § 5.3, the Pauli principle and its particular expression, the Pauli exclusion principle, will be stated; the theorem of spin-statistics connection will allow one to recognize which type of statistics is applicable according to the nature of the considered particles and to distinguish between fermions and bosons. In § 5.4, the special case of two identical particles of spin 1/2 will be analyzed. These results will be extended to an arbitrary number of particles in § 5.5 and the description of a N indistinguishable particle state through the occupation numbers of its various energy levels will be introduced : this is much more convenient than using the N-particle wave function. It is this approach that will be used in the following chapters of this book. Finally in § 5.6 we will return to the examples of the introduction and will interpret them in the framework of the Pauli principle.

5.1   Introduction

It is possible to distinguish fixed particles of the same nature, by specifying their coordinates : this is the case of magnetic moments localized on sites of a paramagnetic crystal, of hydrogen molecules adsorbed on a solid catalyst. They are then *distinguishable* through their coordinates.

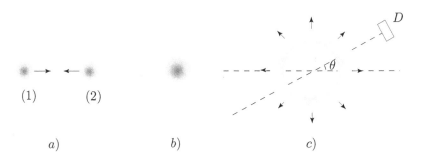

FIG. 5.1: Two identical particles, *a)* represented by the wave packets which describe them in their center-of-mass frame, *b)* suffer a collision. *c)* After the collision, the wave function is spread in a ring-shaped region the radius of which increases with time. When a particle is detected in D, it is impossible to know whether it was associated with wave packet (1) or (2) prior to the collision.

On the other hand, two electrons in motion in a metal, two protons in a nucleus, or more generally all particles of the same chemical nature, *identical*, in motion and thus likely to collide, cannot be distinguished by their physical properties : these identical particles are said to be *indistinguishable*. Indeed, as explained in Quantum Mechanics courses, if these particles, the evolution of which is described through Quantum Mechanics, do collide, their wave functions overlap (fig. 5.1). Then, after the collision it is impossible to know whether (1′) originated from (1) or from (2), since the notion of trajectory does not exist in Quantum Mechanics (Fig. 5.2).

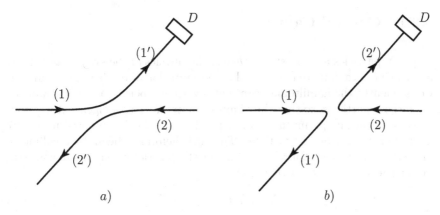

FIG. 5.2: Representation in term of "trajectory" of the event described on Fig. 5.1. The two particles being identical, one cannot distinguish between a) and b).

The quantum treatment of such systems refers to the Pauli principle. This principle allows the interpretation of a great variety of chemical and physical phenomena, among which are

1. the periodic classification of the elements;

2. the chemical binding;

3. the ferromagnetism of some metals, like iron or cobalt, that is, their ability to evidence a macroscopic magnetization even in the absence of an external magnetic field;

4. the superconductivity of some solids, the electrical resistance of which strictly vanishes below a temperature called "critical temperature";

5. the special behavior of the isotope $^4_2$He, which becomes superfluid for temperatures lower than a critical temperature T_c, equal to 2.17 K under the atmospheric pressure : this means that no viscosity is restraining its motion; the isotope $^3_2$He does not show such a property;

6. the stimulated light emission and the laser effect.

5.2   States of Two Indistinguishable Particles (Quantum Mechanics)

5.2.1   General Case

When N particles are *indistinguishable*, the physics of the system and the N-particle hamiltonian are unchanged by permutation of some of these particles. Consequently, the hamiltonian commutes with any permutation operator between these particles, so that they have a common basis of eigenvectors. It can be shown that any permutation of N particles can be decomposed into a product of exchanges of two particles. Thus it is helpful to begin by recalling the properties of the operator which exchanges the particles of a system consisting in particle 1 at $\vec{r}_1$ and particle 2 at $\vec{r}_2$:

$$\hat{P}_{12}\psi(\vec{r}_1, \vec{r}_2) = \psi(\vec{r}_2, \vec{r}_1) \tag{5.1}$$

The operator $\hat{P}_{12}$ must satisfy :

$$(\hat{P}_{12})^2 = 1 \tag{5.2}$$

since after two exchanges one returns to the initial state. This implies that the eigenvalue $e^{i\alpha}$ of $\hat{P}_{12}$ must fulfill the condition :

$$e^{2i\alpha} = 1, \text{ i.e., } e^{i\alpha} = \pm 1 \tag{5.3}$$

i) If

$$e^{i\alpha} = +1, \psi(\vec{r}_2, \vec{r}_1) = \psi(\vec{r}_1, \vec{r}_2) \tag{5.4}$$

the eigenstate is symmetrical in the exchange (or transposition) of particles 1 and 2.

ii) If

$$e^{i\alpha} = -1, \psi(\vec{r}_2, \vec{r}_1) = -\psi(\vec{r}_1, \vec{r}_2) \tag{5.5}$$

the wave function is antisymmetrical in the transposition of the two particles.

5.2.2   Independent Particles

If the indistinguishable particles are *independent*, the total hamiltonian for all the particles is written as a sum of one-particle hamiltonians :

$$\hat{H}_N = \sum_{i=1}^{N} \hat{h}_i \tag{5.6}$$

The eigenstates of each one-particle hamiltonian satisfy :

$$\hat{h}_i \chi_n(\vec{r}_i) = \varepsilon_n \chi_n(\vec{r}_i) \, (i = 1, 2) \tag{5.7}$$

The index n corresponds to the considered quantum state, for example for the electron of a hydrogen atom $n = 1s, 2s, 2p \ldots$; the particle i is located at $\vec{r}_i$.

The *one-particle* eigenstate at the lowest energy (fundamental state) satisfies

$$\hat{h}_1 \chi_1(\vec{r}_1) = \varepsilon_1 \chi_1(\vec{r}_1) \tag{5.8}$$

The energy ε_2 of the *one-particle first* excited state is such that

$$\hat{h}_1 \chi_2(\vec{r}_1) = \varepsilon_2 \chi_2(\vec{r}_1) \tag{5.9}$$

Since the particles are assumed to be independent, the eigenstates of the two-particle hamiltonian $\hat{H}_{12}$ can be easily expressed with respect to the one-particle solutions :

$$\hat{H}_{12} \psi_{nn'}(\vec{r}_1, \vec{r}_2) = \varepsilon_{nn'} \psi_{nn'}(\vec{r}_1, \vec{r}_2) \tag{5.10}$$

with

$$\varepsilon_{nn'} = \varepsilon_n + \varepsilon_{n'} \text{ and } \psi_{nn'}(\vec{r}_1, \vec{r}_2) = \chi_n(\vec{r}_n) \chi_{n'}(\vec{r}_2) \tag{5.11}$$

Indeed remember that

$$\begin{aligned}
(\hat{h}_1 + \hat{h}_2)\chi_n(\vec{r}_1)\chi_{n'}(\vec{r}_2) &= \varepsilon_n \chi_n(\vec{r}_1)\chi_{n'}(\vec{r}_2) + \chi_n(\vec{r}_1)\varepsilon_{n'}\chi_{n'}(\vec{r}_2) \\
&= (\varepsilon_n + \varepsilon_{n'})\chi_n(\vec{r}_1)\chi_{n'}(\vec{r}_2)
\end{aligned} \tag{5.12}$$

so that the two-particle energy is the sum of the one-particle energies and the two-particle wave function is the product of the one-particle wave functions. Any linear combination of wave functions corresponding to the same energy $\varepsilon_n + \varepsilon_{n'}$ is also a solution of $\hat{H}_{12}$.

Here one is looking for eigenfunctions with a symmetry determined by the indistinguishability properties.

In the *two-particle fundamental* state, each individual particle is in its fundamental state (fig. 5.3), this two-particle state has the energy $2\varepsilon_1$, it is nondegenerate. Its wave function satisfies

$$\hat{H}_{12} \psi_{11\text{sym}}(\vec{r}_1, \vec{r}_2) = 2\varepsilon_1 \psi_{11\text{sym}}(\vec{r}_1, \vec{r}_2) \tag{5.13}$$

It is symmetrical and can be written

$$\psi_{11\text{sym}}(\vec{r}_1, \vec{r}_2) = \chi_1(\vec{r}_1)\chi_1(\vec{r}_2) . \tag{5.14}$$

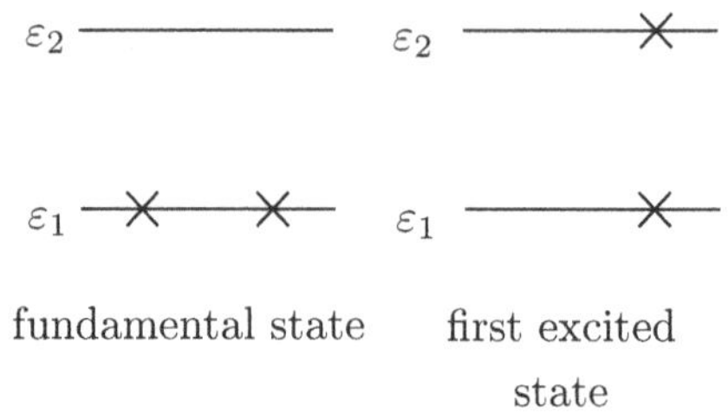

ε_2 —————— ε_2 ———✗——

ε_1 ✗———✗— ε_1 ————✗——

fundamental state   first excited
state

FIG. 5.3 : Occupation of two levels by two indistinguishable particles.

In the *two-particle first excited state*, the total energy is $\varepsilon_1 + \varepsilon_2$. This state is doubly degenerate, as two distinct independent states can be built (particle 1 in ε_1, particle 2 in ε_2 or the opposite). But since the particles are indistinguishable, one cannot tell which of the two particles is in the fundamental state. Thus one looks for solutions which are common eigenstates of $\hat{H}_{12}$ and $\hat{P}_{12}$. The acceptable solutions are :

for the symmetrical first excited state

$$\psi_{12\text{sym}}(\vec{r}_1, \vec{r}_2) = \frac{1}{\sqrt{2}}(\chi_1(\vec{r}_1)\chi_2(\vec{r}_2) + \chi_2(\vec{r}_1)\chi_1(\vec{r}_2)) \qquad (5.15)$$

for the antisymmetrical first excited state

$$\psi_{12\text{asym}}(\vec{r}_1, \vec{r}_2) = \frac{1}{\sqrt{2}}(\chi_1(\vec{r}_1)\chi_2(\vec{r}_2) - \chi_2(\vec{r}_1)\chi_1(\vec{r}_2)) \qquad (5.16)$$

These wave functions (5.14) to (5.16) are determined, to a phase factor.

Now consider the case of N particles.

The total energy of the system, which is the eigenvalue of $\hat{H}_N$, is then equal to

$$\varepsilon_N = \sum_{i=1}^{N} \varepsilon_{n_i} \qquad (5.17)$$

The eigenfunction is a product of one-particle eigenfunctions,

$$\psi_N(\vec{r}_1, \ldots, \vec{r}_N) = \prod_{i=1}^{N} \chi_{n_i}(\vec{r}_i) \qquad (5.18)$$

Any permutation of N particles can be decomposed into a product of exchanges of particles by pair, it thus admits the eigenvalues $+1$ or -1 :

$$\hat{P}\psi(\vec{r}_1, \vec{r}_2, \ldots \vec{r}_i, \ldots \vec{r}_N) = \sigma\psi(\vec{r}_1, \vec{r}_2, \ldots \vec{r}_i, \ldots, \vec{r}_N) \qquad (5.19)$$

with $\sigma = \pm 1$.

$\sigma = +1$ is associated with an "even" permutation

$\sigma = -1$ is associated with an "odd" permutation

Although the decomposition of the permutation into a product of exchanges is not unique, the value of σ is well determined.

Among the eigenfunctions of the permutation operators, two types will be distinguished : either the wave function is *completely symmetrical*, i.e., it is unchanged by any permutation operator ; or it is a *completely antisymmetrical* eigenfunction which satisfies (5.19) and changes its sign for an odd permutation, but is not changed for an even permutation. The eigenfunctions, common to the permutation operators and to the hamiltonian $\hat{H}_N$ of the N indistinguishable particles, will necessarily be of one of these two types.

5.3   Pauli Principle ; Spin-Statistics Connection

5.3.1   Pauli Principle ; Pauli Exclusion Principle

The Pauli (1900-1958) principle is a postulate which is stated as follows :

The wave function of a set of N indistinguishable particles obeys either of the two properties below :

– either it is completely symmetrical, i.e., its sign does not change by permutation of two arbitrary particles : then these particles are called **bosons***;*

– or it is completely antisymmetrical, and thus its sign is modified by permutation of two particles : these particles are **fermions***.*

The property verified by the wave function only depends on the type of the considered particles, it is independent of their number N and of the particular state Ψ they occupy.

This principle is verified by all chemical types of particles. In particular, every time a new type of elementary particle was discovered, these new particles indeed satisfied the Pauli principle.

A special case of the Pauli principle is the Pauli exclusion principle : *Two fermions cannot be in the same quantum state.* Indeed, if they were in the

same one-particle state Ψ, the antisymmetrical expression of the wave function
would be :

$$\psi(1,2) = \frac{1}{\sqrt{2}}[\psi(1)\psi(2) - \psi(2)\psi(1)] = 0 \qquad (5.20)$$

One now needs to determine which particles, called "fermions", satisfy the
Fermi-Dirac statistics, and which particles, the "bosons", follow the Bose-
Einstein statistics. There is no intuitive justification to the following theorem,
which provides the answer to this question.

5.3.2   Theorem of Spin-Statistics Connection

> *The particles with an integral or zero spin $(S/\hbar = 0, 1, 2\ldots)$ are*
> *bosons. The wave function for N bosons is symmetrical in the*
> *transposition of two of them.*
> *The particles with an half-integral spin $(S/\hbar = 1/2, 3/2, 5/2\ldots)$*
> *are fermions. The N-fermions wave function is antisymmetrical*
> *in the transposition of two of them.*

Among the elementary particles, electrons, protons, neutrons, neutrinos,
muons have a spin equal to $1/2$; the spin of photons is 1, the spin of me-
sons is 0 or 1.

When these particles are combined into one or several nuclei, the total angular
momentum takes into account all the orbital momenta and the spin intrinsic
momenta. In Quantum Mechanics courses it is shown that the addition of two
angular momenta J_1 and J_2 gives a parameter which has the properties of a
new angular momentum, with a quantum number J varying between $|J_1 - J_2|$
and $|J_1 + J_2|$ by unity steps. If two integral or half-integral momenta are
added, only integral values of J are obtained; if an integral J_1 is added to a
half-integral J_2, the sum J is half-integral. It is the quantum number J which
is identified with the spin of the ensemble.

More generally if a composite particle contains an even number of elementary
particles of half-integral spin, its total angular moment (or its total spin)
is integral, this is a boson; if it contains an odd number, its total angular
moment is half-integral, and the composite particle is a fermion.

For example, the atom $^4_2$He has two electrons, two protons and two neutrons,
i.e., six particles of half-integral spin : it is a boson. On the other hand, the
atom $^3_2$He is composed of two electrons, two protons but a single neutron,
thus it is a fermion. The statistical properties of these two isotopes are very
different (see the Introduction of this chapter, example 5), whereas the che-

mical properties, determined by the number of electrons that can participate in chemical bonds, are the same.

5.4 Case of Two Particles of Spin 1/2

Here the properties of the wave functions of two indistinguishable particles of spin 1/2 are recalled, this is in particular the case for two electrons. From the theorem of spin-statistics connection, this wave function must be anti-symmetrical with respect to the transposition of these two particles. Now, in addition to their spin degree of freedom, these two particles also have an orbital degree of freedom concerning the space coordinates; from Quantum Mechanics courses, the wave function for a single particle of this type can be decomposed into tensor products concerning the space and spin variables. The most general form of decomposition is :

$$\psi(\vec{r}_1, \vec{r}_2; \sigma_1, \sigma_2) = \psi_{++}(\vec{r}_1, \vec{r}_2)| + + \rangle + \psi_{+-}(\vec{r}_1, \vec{r}_2)| + - \rangle$$
$$+ \psi_{-+}(\vec{r}_1, \vec{r}_2)| - + \rangle + \psi_{--}(\vec{r}_1, \vec{r}_2)| - - \rangle \qquad (5.21)$$

5.4.1 Triplet and Singlet Spin States

Let us first assume, for simplification, that this decomposition is reduced to a single term, i.e.,

$$\psi(\vec{r}_1, \vec{r}_2; \sigma_1, \sigma_2) = \psi(\vec{r}_1, \vec{r}_2) \otimes |\sigma_1, \sigma_2\rangle, \text{ with } \sigma_1, \sigma_2 = \pm 1/2 \qquad (5.22)$$

Since we are dealing with fermions, the two-particle wave function changes sign when particles 1 and 2 are exchanged. This sign change can occur in two different ways :

– either it is the space part $\psi(\vec{r}_1, \vec{r}_2)$ of the wave function which is antisymmetrical in this exchange, the spin function remaining unchanged ;

– or it is $|\sigma_1, \sigma_2\rangle$ which changes its sign in this exchange, the space function remaining unchanged.

Consider the spin states. Each particle has two possible spin states, noted $| \uparrow \rangle$ or $| \downarrow \rangle$. There are thus four basis states for the set of two spins, which can be chosen as

$$| \uparrow\uparrow \rangle, | \uparrow\downarrow \rangle, | \downarrow\uparrow \rangle, | \downarrow\downarrow \rangle \qquad (5.23)$$

(the order of notation is : particle 1, then particle 2).

From these states one deduces four new independent states, classified according to their symmetry, that is, their behavior when exchanging the spins of
the two particles.

There are three symmetrical states, unchanged when the role of particles 1
and 2 is exchanged :

$$\left\{ \begin{array}{l} |\uparrow\uparrow\rangle = |11\rangle \\[2mm] \dfrac{1}{\sqrt{2}}(|\uparrow\downarrow\rangle + |\downarrow\uparrow\rangle) = |10\rangle \\[2mm] |\downarrow\downarrow\rangle = |1-1\rangle \end{array} \right. \qquad (5.24)$$

They constitute the *triplet* state of total spin $S = 1$, with projections
on the quantization axis Oz described by the respective quantum numbers
$S_z = +1, 0, -1$: the eigenvalue of the squared length of this spin is equal to
$S(S+1)\hbar^2$, i.e., $2\hbar^2$ when the eigenvalue of the projection along Oz is $S_z\hbar$,
i.e., $\hbar, 0$ or $-\hbar$.

There remains an antisymmetrical state, which changes its sign in the exchange of the two spins, this is the *singlet state*

$$\frac{1}{\sqrt{2}}(|\uparrow\downarrow\rangle - |\downarrow\uparrow\rangle) = |00\rangle \qquad (5.25)$$

in which $S = S_z = 0$.

Even when the two-particle hamiltonian does not explicitly depend on spin,
orbital wave functions with different symmetries and possibly different energy
eigenvalues are associated with the triplet and singlet states : for example, in
the special case treated in § 5.2.2 the fundamental state, with a symmetrical
orbital wave function [see Eq. (5.14)], must be a spin singlet state, whereas
in the first excited state one can have a spin triplet. This energy splitting
between the symmetrical and antisymmetrical orbital solutions is called the
exchange energy or *exchange interaction*.

The exchange interaction is responsible of ferromagnetism, as understood by
Werner Heisenberg (1901-1976) as early as 1926 : indeed the dipolar interaction between two magnetic moments a fraction of a nm apart, each of the
order of the Bohr magneton, is much too weak to allow the existence of an
ordered phase of magnetization at room temperature. On the contrary, owing
to the Pauli principle, in ferromagnetic systems the energy splittings between
states with different spins correspond to separations between different orbital
levels. These are of a fraction of an electron-volt, which is equivalent to temperatures of the order of 1000 K, since these states are the ones responsible
for valence bindings.

In a ferromagnetic solid, magnetism can originate from mobile electrons in metals like in Fe, Ni, Co, or from ions in insulating solids, as in the magnetite iron oxide Fe_3O_4, and the spin can differ from $1/2$. However this interaction between the spins, which relies on the Pauli principle, is responsible for the energy splitting between different orbital states : one state in which the total magnetic moment is different from zero, corresponding to the lower temperature state, and the other one in which it is zero.

5.4.2 General Properties of the Wave Function of Two Spin $1/2$ Particles

It has just been shown that, because of the Pauli principle, space wave functions are associated with the triplet states, which are changed into their opposite in the exchange of the two particles (antisymmetrical wave function). To the singlet state is associated a symmetrical wave function. The most general wave function for two spin 1/2 particles will be written, for a decomposition onto the triplet and singlet states :

$$\psi_{11}^A|11\rangle + \psi_{10}^A|10\rangle + \psi_{1-1}^A|1-1\rangle + \psi_{00}^S|00\rangle$$

The wave functions of the type ψ_{11}^A are spatial functions of $\vec{r}_1$ and $\vec{r}_2$, the upper index recalls the symmetry of each component.

The same wave function can be expressed on the spin basis $|\sigma_1, \sigma_2\rangle$, accounting for expressions (5.24) and (5.25) of the triplet and singlet wave functions. One thus obtains :

$$\psi(\vec{r}, \vec{r}_2; \sigma_1, \sigma_2) = \psi_{11}^A|++\rangle + \frac{\psi_{10}^A + \psi_{00}^S}{\sqrt{2}}|+-\rangle + \frac{\psi_{10}^A - \psi_{00}^S}{\sqrt{2}}|-+\rangle$$
$$+ \psi_{1-1}^A|--\rangle \tag{5.26}$$

It thus appears, by comparing to the general expression (5.21), that the Pauli principle dictates symmetry conditions on the spin 1/2 components :

- $\psi_{++}(\vec{r}_1, \vec{r}_2)$ and $\psi_{--}(\vec{r}_1, \vec{r}_2)$ must be antisymmetrical in the exchange of particles 1 and 2

- $\psi_{+-}(\vec{r}_1, \vec{r}_2)$ and $\psi_{-+}(\vec{r}_1, \vec{r}_2)$ do not have a specific symmetry, but their sum is antisymmetrical and their difference symmetrical.

5.5 Special Case of N Independent Particles; Occupation Numbers of the States

In the general case of an N-identical-particle system, the Pauli principle dictates a form for the N-particle wave function and occupation conditions for the energy levels which are very different in the case of fermions or bosons. In this part of the course, we are only interested by independent particles : this situation is simpler and already allows a large number of physical problems to be solved.

5.5.1 Wave Function

Fermions :

We have seen that two fermions cannot be in the same quantum state and that their wave function must be antisymmetrical in the transposition of the two particles. We consider independent particles, the N-particle wave function of which takes the form of a determinant called "Slater determinant," which indeed ensures the change of sign by transposition and the nullity if two particles are in the same state :

$$\psi(\vec{r}_1, \ldots, \vec{r}_N) \propto \begin{vmatrix} \chi_1(\vec{r}_1) & \chi_2(\vec{r}_1) & \chi_N(\vec{r}_1) \\ \vdots & \vdots & \vdots \\ \chi_1(\vec{r}_N) & \vdots & \chi_N(\vec{r}_N) \end{vmatrix} \quad (5.27)$$

If a state is already occupied by a fermion, another one cannot be added. A consequence is that at zero temperature all the particles cannot be in the minimum energy state : the different accessible levels must be filled, beginning with the lowest one, each time putting one particle per level and thus each time moving to the next higher energy level. The process stops when the Nth and last particle is placed in the last occupied level : in Chemistry this process is called the "Aufbau principle" or construction principle; in solids, the last occupied state at zero temperature is called the Fermi level and is usually noted ε_F (see chapters 7 and 8).

Bosons :

On the other hand, the wave function for N bosons is symmetrical; in particular it is possible to have all the particles on the same level of minimum energy at zero temperature.

5.5.2   Occupation Numbers

The N-fermion (or N-boson) wave functions are complicated to write because, as soon as a particle is put in a state of given energy, this induces conditions on the available states for the other particles ; then the antisymmetrization of the total wave function has to be realized for the fermions (or the symmetrization for the bosons). Since these particles are indistinguishable, the only interesting datum in Statistical Physics is the number of particles in a defined quantum state.

Instead of writing the N-particle wave function, which will be of no use in what follows, we will described the system as a whole by the occupation numbers of its different states, this is the so-called "Fock basis" : these one-particle states correspond to distinct ε_k, which can have the same energy but different indexes if the state is degenerate (for example two states of the same energy and different spins will have different indexes k and k') ; the states are ordered by increasing energies (Fig. 5.4).

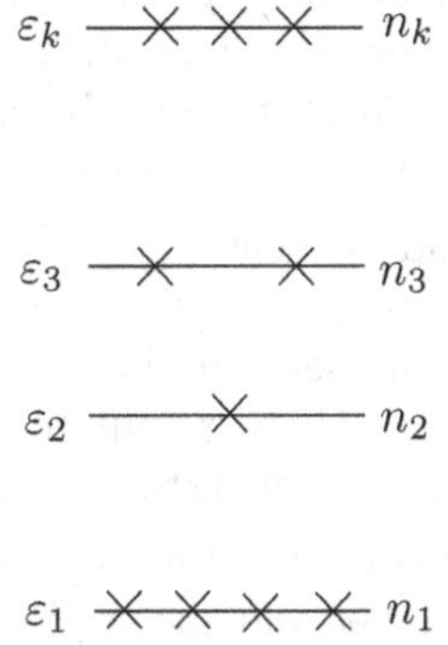

FIG. 5.4 : Definition of the occupation numbers of the one-particle states.

The notation $|n_{1,2}, \ldots n_k, \ldots\rangle$ will express the N-particle state in which n_1 particles occupy the state of energy ε_1, n_2 that of energy $\varepsilon_2, \ldots n_k$ the state of energy ε_k. One must satisfy

$$n_1 + n_2 + \ldots + n_k + \ldots = N \tag{5.28}$$

the total number of particles in the system. Under these conditions, the total energy of the N-particle system is given by

$$n_1\varepsilon_1 + n_2\varepsilon_2 + \ldots + n_k\varepsilon_k + \ldots = E_N \tag{5.29}$$

From the Pauli principle

- for fermions : $n_k = 0$ or 1 $\tag{5.30}$

- for bosons   : $n_k = 0, 1, 2 \ldots \infty$ $\tag{5.31}$

This description will be used in the remainder of the course. The whole physical information is included in the Fock notation. Indeed one cannot tell which particle is in which state, this would be meaningless since the particles are indistinguishable. The only thing that can be specified is how many particles are in each one-particle state.

5.6 Return to the Introduction Examples

Now we are able to propose a preliminary interpretation of the physical phenomena described in the introduction, § 5.1, some of which will be analyzed in more detail in the following chapters :

5.6.1 Fermions Properties

1. In the periodic classification of the elements, the atomic levels are filled with electrons which are *fermions*. In a given orbital level only two electrons of distinct spin states can be placed, i.e., according to the Pauli principle they are in the singlet spin state.

2. When a chemical bond is created, the electrons coming from the two atoms of the bond go into the atomic levels originating from the coupling between the levels of the separated atoms. The resulting bonding and antibonding states must also be populated by electrons in different spin states, thus in the singlet spin state.

3. The ferromagnetism arises, as already mentioned, from the energy difference between the singlet and the triplet spin states for a pair of electrons. If the fundamental state presents a macroscopic magnetization, it must be made from microscopic states with a nonzero spin. In fact, in a ferromagnetic metal such as iron or nickel, the spins of two iron nuclei couple through conduction electrons which travel from nucleus to nucleus.

5.6.2 Bosons Properties

4. In superconductivity the electrons of a solid couple by pairs due to their interaction with the solid vibrations. These pairs, called "Cooper pairs," are bosons which gather into the same fundamental state at low temperature, thus producing the phenomenon of superconductivity.

5. In the same way the ^{4}He atoms are bosons, which can occupy the same quantum state below a critical temperature (see chapter 9). Their ma-

croscopic behavior expresses this collective property, whence the absence of viscosity.

6. Photons are also bosons and the stimulated emission, a specific property of bosons in nonconserved number, will be studied in § 9.2.4.

Summary of Chapter 5

In Quantum Mechanics, the hamiltonian describing the states of identical particles is invariant under a permutation of these particles, the states before and after permutation are physically indistinguishable. This means that the hamiltonian commutes with any permutation operator : one first looks for the permutations eigenstates, which are also eigenstates of the N-particle hamiltonian.

In the case of two indistinguishable particles the symmetrical and antisymmetrical eigenstates have been described. These states have a simple expression in the case of independent particles.

Any permutation of N particles can be decomposed into a product of exchanges (or transpositions) of particles by pair : a permutation is thus either even or odd.

The Pauli principle postulates that the wave functions for N identical particles can only be, according to the particle's nature, either completely symmetrical (bosons) or completely antisymmetrical (fermions) in a permutation of the particles. In particular, because their wave function is antisymmetrical, two fermions cannot occupy the same quantum state ("Pauli exclusion principle").

The theorem of spin-statistics connection specifies that the particles of half-integral total spin, the fermions, have an antisymmetrical wave function ; the particles of integral or zero total spin, the bosons, have a completely symmetrical wave function.

In the case of two spin-1/2 particles, the triplet spin state, symmetrical, and the singlet spin state, antisymmetrical, have been described. The spin component of the wave function has to be combined with the orbital part, the global wave function being antisymmetrical to satisfy the Pauli principle.

In Statistical Physics, to account for the Pauli principle, the state of N particles will be described using the occupation numbers of the various energy levels, rather than by its wave function. A given state can be occupied at most by one fermion, whereas the number of bosons that can be located in a given state is arbitrary.

Chapter 6

General Properties of the Quantum Statistics

In chapter 5 we learnt the consequences, in Quantum Mechanics, of the Pauli principle which applies to indistinguishable particles : they will now always be taken as independent, that is, without any mutual interaction.

We saw that it is very difficult to express the conditions stated by the Pauli principle on the N-particle wave function; on the other hand, a description through occupation numbers $|n_1, n_2, \ldots, n_k, \ldots\rangle$ of the different quantum states $\varepsilon_1, \varepsilon_2, \ldots \varepsilon_k \ldots$ accessible to these indistinguishable particles is much more convenient. In fact we assume that the states are nondegenerate and if there are several states with the same energy, we label them with different indexes. Besides, it was stated that there exists two types of particles, the fermions with a half-integral spin, which cannot be more than one in any given one-particle state ε_k, whereas the bosons, with an integral or zero spin, can be in arbitrary number in any one single-particle quantum state.

In the present chapter we return to Statistical Physics (except in § 6.4, which is on Quantum Mechanics) : after having presented the study technique applicable to systems referring to Quantum Statistics, we determine the average number of particles, *at a given temperature*, on an energy state of the considered system : this number will be very different for fermions and bosons.

This chapter presents general methods and techniques to analyze, using Statistical Physics, the properties of systems of indistinguishable particles following the Quantum Statistics of either Fermi-Dirac or Bose-Einstein, while the following chapters will study these statistics more in detail on physical examples.

In §6.1, we will show how the statistical properties of indistinguishable independent particles are more conveniently described in the grand canonical ensemble. In §6.2 we will explain how the grand canonical partition function Z_G is factorized into terms, each one corresponding to a one-particle quantum state. In §6.3 the Fermi-Dirac and the Bose-Einstein distributions will be deduced, which give the average number of particles of the system occupying a quantum state of given energy at temperature T, and the associated thermodynamical potential, the grand potential, will be expressed.

The §6.4 will be a (long) Quantum Mechanics parenthesis, it will be shown there that a macroscopic system, with extremely close energy levels, is conveniently described by a density of states, that will be calculated for a free particle. In §6.5, this density of states will be used to obtain the average value, at fixed temperature, of physical parameters in the case of indistinguishable particles. Finally, in §6.6, the criterion used in §4.3 to define the domain of Classical Statistics will be justified and it will be shown that the Maxwell-Boltzmann Classical Statistics is the common limit of both Quantum Statistics when the density decreases or the temperature increases.

6.1   Use of the Grand Canonical Ensemble

In chapter 5 we learnt the properties requested for the wave function of N *indistinguishable* particles, when these particles are fermions or when they are bosons. (From the present chapter until the end of the book, the study will be limited to indistinguishable and *independent* particles.) Now we look for the statistical description of such a system in thermal equilibrium at temperature T and first show, on the example of two particles, that the use of the canonical ensemble is not convenient.

6.1.1   Two Indistinguishable Particles in Thermal Equilibrium : Statistical Description in the Canonical Ensemble

Consider a system of two indistinguishable particles. The one-particle energy levels are noted $\varepsilon_1, \ldots, \varepsilon_k, \ldots$, and are assumed to be nondegenerate (a degenerate level is described by several states at the same energy but with different indexes). The canonical partition function for a *single* particle at temperature T is written :

$$Z_1 = \sum_k e^{-\beta \varepsilon_k} \quad \text{with } \beta = \frac{1}{k_B T} \tag{6.1}$$

Now consider the two particles. For a given microscopic configuration, the total energy is $\varepsilon_i + \varepsilon_j$. After having recalled the results for two *distinguishable* particles, we will have to separately consider the case of *fermions* and that of *bosons*.

We have seen in § 2.4.5 that for *distinguishable* particles

$$Z_{2\text{disc}} = (Z_1)^2 \tag{6.2}$$

since there is no limitation to the levels occupation by the second particle.

Fermions cannot be more than one to occupy the same state. Consequently, as soon as the level of energy ε_i is occupied by the first particle, the second one cannot be on it. Thus

$$Z_{2\text{fermions}} = \frac{1}{2} \sum_{i \neq j} e^{-\beta(\varepsilon_i + \varepsilon_j)} = \sum_{i < j} e^{-\beta(\varepsilon_i + \varepsilon_j)} \tag{6.3}$$

Let us compare $Z_{2\text{fermions}}$ to $(Z_1)^2$:

$$(Z_1)^2 = \left(\sum_i e^{-\beta \varepsilon_i} \right) \left(\sum_j e^{-\beta \varepsilon_j} \right) = \sum_i e^{-\beta 2\varepsilon_i} + 2 \sum_{i < j} e^{-\beta(\varepsilon_i + \varepsilon_j)} \tag{6.4}$$

$$(Z_1)^2 = \sum_i e^{-\beta 2\varepsilon_i} + 2 Z_{2\text{fermions}} \tag{6.5}$$

In the case of *bosons*, in addition to the configurations possible for the fermions, the two particles can also lie on the same level :

$$Z_{2\text{bosons}} = \sum_{i \leq j} e^{-\beta(\varepsilon_i + \varepsilon_j)} \tag{6.6}$$

Let us relate $(Z_1)^2$ to $Z_{2\text{bosons}}$:

$$(Z_1)^2 = Z_{2\text{bosons}} + \frac{1}{2} \sum_{i \neq j} e^{-\beta(\varepsilon_i + \varepsilon_j)} \tag{6.7}$$

$$(Z_1)^2 = Z_{2\text{bosons}} + Z_{2\text{fermions}} \tag{6.8}$$

It was easy to determine the levels available for the particles in this example because they were only two. For three particles one must then consider which restrictions appear for the third one because of the levels occupied by the first two particles. For N particles one should again proceed step by step, but this method is no longer realistic !

6.1.2 Description in the Grand Canonical Ensemble

To avoid complicated conditions on the occupation numbers n_k one will work in the *grand canonical ensemble*, assuming *in a first step* that the *total number of particles* of the system is *arbitrary*. Thus one will calculate Z_G and the grand potential A; then the *value* of the Lagrange parameter $\alpha = \beta\mu$ (or that of the *chemical potential* μ) will be fixed in such a way that the *average number of particles* $\langle N \rangle$, deduced from the statistical calculation, *coincides* with the *real number* of particles in the system. We know (§ 2.5.4) that the relative fluctuation on N introduced by this procedure is of the order of $1/\sqrt{N}$, that is, of the order of a few 10^{-12} for N of the order of the Avogadro number.

6.2   Factorization of the Grand Partition Function

6.2.1   Fermions and Bosons

By definition

$$Z_G(\alpha, \beta) = \sum_{N=0}^{\infty} e^{\alpha N} Z_N(\beta) = \sum_{N=0}^{\infty} e^{\beta\mu N} Z_N(\beta) \tag{6.9}$$

where N is one of the numbers of particles and where α and the chemical potential are related by $\alpha = \beta\mu$. The canonical partition function $Z_N(\beta)$ corresponding to N particles is given by

$$Z_N(\beta) = \sum_{n} e^{-\beta E_n} \tag{6.10}$$

Let us introduce the occupation numbers n_k of the quantum states ε_k :

$$Z_G = \sum_{N=0}^{\infty} e^{\beta\mu \sum_{k} n_k} \left(\sum_{\sum n_k = N} e^{-\beta \sum_{k} n_k \varepsilon_k} \right) \tag{6.11}$$

with

$$\sum_{k} n_k = N \tag{6.12}$$

and the energy for this configuration of the N particles being given by

$$\sum_{k} n_k \varepsilon_k = E_N \tag{6.13}$$

As the sum is performed over all the total numbers N and all the particle distributions in the system, Z_G is also written

$$Z_G = \sum_{\{|n_1...n_k>\}} e^{\beta\mu\sum_k n_k - \beta\sum_k n_k\varepsilon_k} \tag{6.14}$$

$$= \sum_{\{|n_1...n_k>\}} e^{\beta\sum_k n_k(\mu-\varepsilon_k)} \tag{6.15}$$

In the latter expression, one separates the contribution of the state ε_1 : this is the term

$$\sum_{n_1} e^{\beta(\mu-\varepsilon_1)n_1} \tag{6.16}$$

which multiplies the sum concerning all the other states, with n_1 taking all possible values. One then proceeds state after state, so that Z_G is written as a product of factors :

$$Z_G = \prod_{k=1}^{\infty} \left(\sum_{n_k} e^{\beta(\mu-\varepsilon_k)n_k} \right) \tag{6.17}$$

Each factor concerns a single state ε_k and its value depends on the number of particles which can be located in this state, i.e., of the nature, fermions or bosons, of the considered particles. Note that, since Z_G is a product of factors, $\ln Z_G$ is a sum of terms, each concerning a one-particle state of the type ε_k.

6.2.2 Fermions

In this case, the quantum state ε_k is occupied by *0 or 1 particle* : there are only two terms in $\sum n_k$, the contribution of the state ε_k in Z_G is the factor $1 + \exp\beta(\mu - \varepsilon_k)$, whence

$$Z_{G\text{ fermions}} = \prod_{k=1}^{\infty}(1 + e^{\beta(\mu-\varepsilon_k)}) \tag{6.18}$$

6.2.3 Bosons

The state ε_k can now be occupied by *0, 1, 2,...* *particles*, so that n_k varies from zero to infinity. The contribution in Z_G of the state ε_k is thus the factor

$$\sum_{n_k=0}^{\infty} e^{\beta(\mu-\varepsilon_k)n_k} = \frac{1}{1 - e^{\beta(\mu-\varepsilon_k)}} \tag{6.19}$$

This geometrical series can be summed only if $\exp(\alpha - \beta\varepsilon_k)$ is smaller than unity, a condition to be fulfilled by all the states ε_k. One will thus have to verify that $\mu < \varepsilon_1$, where ε_1 is the one-particle quantum state with the lowest energy, i.e., the fundamental state. This condition is also expressed through $\alpha = \beta\mu$ as $\alpha < \beta\varepsilon_1$.

Then the grand partition function takes the form :

$$Z_{G \text{ bosons}} = \prod_{k=1}^{\infty} \frac{1}{1 - e^{\beta(\mu-\varepsilon_k)}} \tag{6.20}$$

6.2.4   Chemical Potential and Number of Particles

The grand canonical partition function Z_G is associated with the grand potential A such that

$$A = -k_B T \ln Z_G \tag{6.21}$$

As specified in §3.5.3, the partial derivatives of the grand partition function provide the entropy, the pressure and the average number of particles N :

$$dA = -SdT - Pd\Omega - Nd\mu \tag{6.22}$$

In particular the constraint on the total number of particles of the system is expressed from :

$$N = -\left(\frac{\partial A}{\partial \mu}\right)_{T,\Omega} \tag{6.23}$$

6.3   Average Occupation Number ; Grand Potential

From the grand partition function Z_G one deduces average values at temperature T of physical parameters of the whole system, but also of the state ε_k. In particular the average value of the occupation number $\langle n_k \rangle$ of the state of energy ε_k is obtained through

$$\langle n_k \rangle = \frac{\sum_{n_k} n_k \, e^{\beta(\mu-\varepsilon_k)n_k}}{\sum_{n_k} e^{\beta(\mu-\varepsilon_k)n_k}} = -\frac{1}{\beta}\frac{\partial \ln Z_G}{\partial \varepsilon_k} \tag{6.24}$$

For fermions, the average occupation number, the so-called "Fermi-Dirac distribution," is given by

$$\langle n_k \rangle_{FD} = \frac{1}{e^{\beta(\varepsilon_k - \mu)} + 1} = \frac{1}{e^{\beta \varepsilon_k - \alpha} + 1} \qquad \text{with } \alpha = \beta \mu \qquad (6.25)$$

For bosons, this average number is "the Bose-Einstein" distribution which takes the form

$$\langle n_k \rangle_{BE} = \frac{1}{e^{\beta(\varepsilon_k - \mu)} - 1} = \frac{1}{e^{\beta \varepsilon_k - \alpha} - 1} \qquad (6.26)$$

Note : Remember the sign difference in the denominator between fermions and bosons! Its essential physical consequences are developed in the next chapters.

The average occupation numbers are related to factors in the grand partition function Z_G : indeed one will verify that

$$\frac{1}{1 - \langle n_k \rangle_{FD}} = 1 + e^{\beta(\mu - \varepsilon_k)} \qquad (6.27)$$

$$1 + \langle n_k \rangle_{BE} = \frac{1}{1 - e^{\beta(\mu - \varepsilon_k)}} \qquad (6.28)$$

which are the respective contributions of the level ε_k to Z_G, in the cases of fermions or bosons.

These occupation numbers yield the average values, at given temperature, thus at fixed β, of physical parameters.

Thus the average number of particles is related to α and β, which appear in $f(\varepsilon)$, through

$$\langle N \rangle = \sum_k \langle n_k \rangle \qquad (6.29)$$

The total energy of the system at temperature T is obtained from

$$U = \sum_k \varepsilon_k \langle n_k \rangle \qquad (6.30)$$

The grand potential A is expressed, like the Z_G factors, versus the occupation numbers of the states of energy ε_k :

$$A = +k_B T \sum_k \ln(1 - \langle n_k \rangle) \quad \text{fermions} \qquad (6.31)$$

$$A = -k_B T \sum_k \ln(1 + \langle n_k \rangle) \quad \text{bosons} \qquad (6.32)$$

6.4 Free Particle in a Box; Density of States (Quantum Mechanics)

The different physical parameters, like the average number of particles $\langle N \rangle$, the internal energy U, the grand potential A, were expressed above as sums of contributions arising from the discrete one-particle levels. Now we will first recall what these states are for a free particle confined within a volume of macroscopic characteristic dimensions, i.e., of the order of a mm or a cm (§6.4.1) : such dimensions are extremely large with respect to atomic distances, so that the characteristic energy splittings are very small (§6.4.1a). The boundary conditions (§6.4.1b) of the wave function imply quantization conditions, that we are going to express in two different ways. However, for both quantization conditions the same energy density of states $D(\varepsilon)$ is defined (§6.4.2), such that the number of allowed states between energies ε and $\varepsilon + d\varepsilon$ is equal to $D(\varepsilon) \, d\varepsilon$.

6.4.1 Quantum States of a Free Particle in a Box

a) Eigenstates

By definition a free particle feels no potential energy (except the one expressing the possible confinement) We first consider that the free particle can move through the entire space. Its hamiltonian is given by

$$\hat{h} = \frac{\hat{p}^2}{2m} \tag{6.33}$$

We are looking for its stationary states (see a course on Quantum Mechanics) : the eigenstate $|\psi\rangle$ and the corresponding energy ε satisfy

$$\frac{\hat{p}^2}{2m}|\psi\rangle = \varepsilon|\psi\rangle \tag{6.34}$$

i.e.,

$$-\frac{\hbar^2}{2m}\Delta\psi(\vec{r}) = \varepsilon\psi(\vec{r}) \tag{6.35}$$

The time-dependent wave function is then deduced :

$$\Psi(\vec{r},t) = \psi(\vec{r})\exp(-i\varepsilon t/\hbar) \tag{6.36}$$

It is known that the three space variables separate in eq. (6.35) and that it is enough to solve the one-dimension Schroedinger equation

$$-\frac{\hbar^2}{2m}\frac{d^2\psi(x)}{dx^2} = \varepsilon_x\psi(x) \tag{6.37}$$

which admits two equivalent types of solutions :

i) either

$$\psi(x) = \frac{1}{\sqrt{L_x}} \exp ik_x x \qquad (6.38)$$

with $k_x > 0, < 0$ or zero. The time-dependent wave function

$$\Psi(x,t) = \frac{1}{\sqrt{L_x}} \exp i \left(k_x x - \frac{\varepsilon_x t}{\hbar} \right) \qquad (6.39)$$

with $\varepsilon_x = \dfrac{\hbar^2 k_x^2}{2m} = \hbar\omega_x$ is a *progressive* wave : when the time t increases, a state of given phase propagates toward increasing x's for $k_x > 0$, toward decreasing x's for $k_x < 0$; the space wave function is a constant for $k_x = 0$.

ii) or

$$\psi(x) = A \cos k_x x + B \sin k_x x \qquad (6.40)$$

where A and B are constants.

The time-dependent wave function

$$\Psi(x,t) = (A \sin k_x x + B \cos k_x x) \exp \left(-i \frac{\varepsilon_x t}{\hbar} \right) \qquad (6.41)$$

corresponds to the same energy ε_x or the same pulsation ω_x as in (6.39). The solution of progressive-wave type (6.39) is found again if one chooses $A = 1/\sqrt{L_x}, B = iA$. For real coefficients A and B, when changing the space origin in x there is indeed separation of the space and time variables in (6.41). Then a state of given space phase does not propagate in time, such a wave is *stationary*.

At three dimensions the kinetic energy terms along the three coordinates add up in the hamiltonian; the energy is equal to

$$\varepsilon = \varepsilon_x + \varepsilon_y + \varepsilon_z \qquad (6.42)$$

and the wave function is the product of three wave functions, each concerning a single coordinate.

Now we are going to express, through boundary conditions on the wave function, that the particle is confined within the volume Ω. This will yield quantization conditions on the wave vector and the energy.

b) Boundary Conditions

The volume Ω containing the particles has *a priori* an arbitrary shape. However, it can be understood that each particle in motion is most of the time at a large distance from the container walls, so that the volume properties are not sensitive to the surface properties as soon as the volume is large enough. It is shown, and we will admit it, that the Statistical Physics properties of a *macroscopic system* are indeed independent of the shape of the volume Ω. For convenience we will now assume that Ω is a box (L_x, L_y, L_z), with macroscopic dimensions of the order of a mm or a cm.

We consider a one-dimension problem, on a segment of length L_x. The presence probability of the particle is zero outside the interval $[O, L_x]$, in the region where the particle cannot be found. There are two ways to express this condition :

i) Stationary boundary conditions :

To indicate that the particle cannot leave the interval, one assumes that in $x = 0$ and $x = L_x$ potential barriers are present, which are infinitely high and thus impossible to overcome. Thus the wave function vanishes outside the interval $[O, L_x]$ and also at the extremities of the segment, to ensure its continuity at these points; since it vanishes at $x = 0$, it must be of the form $\psi(x) = A \sin k_x x$ (Fig. 6.1). The cancellation at $x = L_x$ yields

$$k_x L_x = n_x \pi \tag{6.43}$$

where n_x is an integer, so that

$$k_x = n_x \frac{\pi}{L_x} \tag{6.44}$$

The wave vector is an integer multiple of $\frac{\pi}{L_x}$, it is quantized.

For such values of k_x, the time-dependent wave function is given by

$$\Psi(x, t) = A \sin k_x x \exp\left(-i \frac{\varepsilon_x t}{\hbar}\right) \tag{6.45}$$

One understands that the physically distinct allowed values are restricted to $n_x > 0$: indeed taking $n_x < 0$ is just changing the sign of the wave function, which does not change the physics of the problem. The value $n_x = 0$ is to be rejected since a wave function has to be normalized to unity and thus cannot cancel everywhere. In the same way k_y and k_z are quantized by the conditions of cancellation of the wave function on the surface : finally the allowed vectors $\vec{k}$ for the particle confined within the volume Ω are of the form

$$\vec{k} = \left(n_x \frac{\pi}{L_x}, n_y \frac{\pi}{L_y}, n_z \frac{\pi}{L_z}\right) \tag{6.46}$$

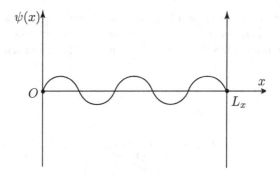

FIG. 6.1: The free-particle wave function is zero outside the interval $]O, L_x[$.

the three integers n_x, n_y, n_z being strictly positive. In the space of the wave vectors (Fig. 6.2), the extremities of these vectors are on a rectangular lattice defined by the three vectors $\left(\dfrac{\pi}{L_x}\vec{i}, \dfrac{\pi}{L_y}\vec{j}, \dfrac{\pi}{L_z}\vec{k}\right)$. The unit cell built on these vectors is $\dfrac{\pi^3}{L_x L_y L_z} = \dfrac{\pi^3}{\Omega}$.

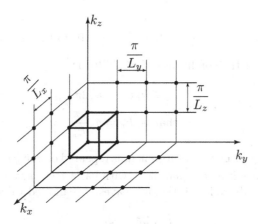

FIG. 6.2: Quantization of the $\vec{k}$-space for boundary conditions of the stationary-wave type. Only the trihedral for which the three $\vec{k}$-components are positive is to be considered.

ii) Periodical boundary [Born-Von Kármán (B-VK)] conditions :

Here again one considers that when a solid is macroscopic, all that deals with its surface has little effect on its macroscopic physical parameters like its pressure, its temperature. The solid will now be closed on itself (Fig. 6.3) and this will not noticeably modify its properties : one thus suppresses surface effects, of little importance for a large system. Then at a given x the properties are the same, and the wave function takes the same value, whether the particle arrives at this point directly or after having traveled one turn around before reaching x :

$$\psi(x + L_x) = \psi(x) \tag{6.47}$$

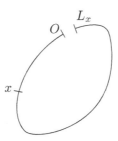

FIG. 6.3 : The segment $[0, L_x]$ is closed on itself.

This condition is expressed on the progressive-wave type wave function (6.39) through

$$\exp ik_x(x + L_x) = \exp ik_x x \tag{6.48}$$

for any x, so that the condition to be fulfilled is

$$k_x L_x = n'_x 2\pi \tag{6.49}$$

$$\text{that is, } k_x = n'_x \frac{2\pi}{L_x} \tag{6.50}$$

where n'_x is integer. The same argument is repeated on the y and z coordinates. The wave vector $\vec{k}$ is again quantized, with basis vectors twice larger than for the condition $i)$ above for stationary waves, the allowed values now being

$$\vec{k} = \left(n'_x \frac{2\pi}{L_x}, n'_y \frac{2\pi}{L_y}, n'_z \frac{2\pi}{L_z} \right) \tag{6.51}$$

The integers n'_x, n'_y, n'_z are now positive, negative or null (Fig. 6.4). Indeed the time-dependent wave function is written, to a phase,

$$\Psi(\vec{r}, t) = \frac{1}{\sqrt{\Omega}} \exp i(\vec{k}.\vec{r} - \omega t), \text{ with } \hbar\omega = \varepsilon = \frac{\hbar^2 k^2}{2m} \tag{6.52}$$

Changing the sign of $\vec{k}$, thus of the integers n'_i, produces a new wave which propagates in the opposite direction and is thus physically different. If one of these integers is zero, this means that the wave is constant along this axis.

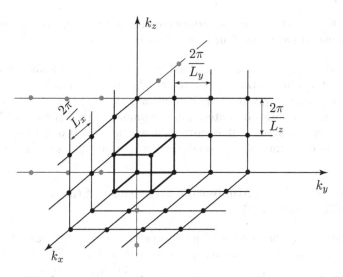

FIG. 6.4: Quantization of the $\vec{k}$-space for the periodic (B-VK) boundary conditions. Note that each dimension of the elementary box is twice that corresponding to the stationary boundary conditions (Fig. 6.2), and that now the entire space should be considered.

6.4.2 Density of States

Under these quantization conditions what is the order of magnitude of the obtained energies, and of the spacing of the allowed levels. As an example consider the framework of the periodic boundary conditions. The quantized allowed energies are given by

$$
\varepsilon = \frac{\hbar^2}{2m} \left[n_x'^2 \left(\frac{2\pi}{L_x} \right)^2 + n_y'^2 \left(\frac{2\pi}{L_y} \right)^2 + n_z'^2 \left(\frac{2\pi}{L_z} \right)^2 \right]
\tag{6.53}
$$

For L_x for the order of 1 mm, a single coordinate and a mass equal to that of the proton,

$$
\frac{\hbar^2}{2m} \left(\frac{2\pi}{L_x} \right)^2 = 13.6 \text{ eV.} \frac{(2\pi)^2}{1840} \cdot \left(\frac{0.5 \times 10^{-10}}{10^{-3}} \right)^2 \cong 8 \times 10^{-16} \text{ eV}
\tag{6.54}
$$

For a mass equal to that of the electron, the obtained value, 1840 times larger, is still extremely small at the electron-volt scale!

The spacing between the considered energy levels is thus very small as compared to any realistic experimental accuracy, as soon as L_x is macroscopic, that is, large with respect to atomic distances (the Bohr orbit of the hydrogen atom is associated with energies in the electron-volt range, see the above estimation). If the stationary boundary conditions had been chosen, the energy splitting would have been four times smaller.

The practical question which is raised is : "What is the number of available quantum states in a wave vector interval $d^3\vec{k}$ around $\vec{k}$ fixed, or in an energy interval $d\varepsilon$ around a given ε ?" It is to answer this question that we are going to now define densities of states, which summarize the Quantum Mechanics properties of the system. It is only in a second step (§ 6.5) that Statistical Physics will come into play through the occupation factors of these accessible states.

a) Wave Vector and Momentum Densities of States (B-VK Conditions) :

One is estimating the number of quantum states dn of a free particle, confined within a volume Ω, of wave vector between $\vec{k}$ and $\vec{k} + d^3\vec{k}$, the volume $d^3\vec{k}$ being assumed to be large as compared to $\dfrac{(2\pi)^3}{\Omega}$. The wave vector density of states $D_{\vec{k}}(\vec{k})$ is defined by

$$dn = D_{\vec{k}}(\vec{k})d^3\vec{k} \tag{6.55}$$

We know that the allowed states are uniformly spread in the $\vec{k}$-space and that, in the framework of the B-VK periodic boundary conditions, for each elementary volume $\dfrac{(2\pi)^3}{\Omega}$ there is a new quantum state (Fig. 6.4) : indeed a parallelepiped has eight summits, shared between eight parallelepipeds. The searched number dn is thus equal to

$$dn = \frac{d^3\vec{k}}{(2\pi)^3/\Omega} = \frac{\Omega}{(2\pi)^3}d^3\vec{k} , \tag{6.56}$$

which determines

$$D_{\vec{k}}(\vec{k}) = \frac{\Omega}{(2\pi)^3} \tag{6.57}$$

The above estimation has been done from the Schroedinger equation solutions related to the space variable $\vec{r}$ (orbital solutions). If the particle carries a spin s, to a given solution correspond $2s+1$ states with different spins, degenerate in the absence of applied magnetic field, which multiplies the above value of the density of states [Eq. (6.57)] by $(2s+1)$. Thus for an electron, of spin $1/2$, there are two distinct spin values for a given orbital state. For a free particle, its wave vector and momentum are related by

$$\vec{p} = \hbar\vec{k} \, d^3\vec{p} = \hbar^3 d^3\vec{k} \tag{6.58}$$

The number dn of states for a free particle of momentum vector between $\vec{p}$ and $\vec{p} + d^3\vec{p}$, confined within the volume Ω, is written, from eq. (6.56),

$$dn = \frac{\Omega}{(2\pi)^3}\frac{d^3\vec{p}}{\hbar^3} = \frac{\Omega}{h^3}d^3\vec{p} = D_{\vec{p}}(\vec{p})d^3\vec{p} \tag{6.59}$$

Here the (three-dimensional) momentum density of states has been defined through

$$D_{\vec{p}}(\vec{p}) = \frac{\Omega}{h^3} \tag{6.60}$$

Note that the distance occupied by a quantum state, on the *momentum* axis of the phase space corresponding to the one-dimension motion of a single particle, is $\dfrac{h}{L_x}$. This is consistent with the fact that a cell of this *phase* space has the area h and that here $\Delta x = L_x$, since, the wave function being of plane-wave type, it is delocalized on the whole accessible distance.

b) Three-Dimensional Energy Density of States (B-VK Conditions) :

The energy of a free particle only contains the kinetic term inside the allowed volume, it only depends of its momentum or wave vector modulus :

$$\varepsilon = \frac{\hbar^2 k^2}{2m} = \frac{p^2}{2m} \tag{6.61}$$

All the states corresponding to a given energy ε are, in the $\vec{k}$-space, at the surface of a sphere of radius $k = \sqrt{\dfrac{2m\varepsilon}{\hbar^2}}$. If one does not account for the particle spin, the number of allowed states between the energies ε and $\varepsilon + d\varepsilon$ is the number of allowed values of $\vec{k}$ between the spheres of radii k and $k + dk$, dk being related to $d\varepsilon$ through $d\varepsilon = \dfrac{\hbar^2}{m}kdk$. (For clarity in the sketch, Fig. 6.5 is drawn in the case of a two-dimensional problem.)

The concerned volume is $4\pi k^2 dk$, the number dn of the searched states, without considering the spin variable, is equal to

$$\frac{\Omega}{(2\pi)^3}4\pi k^2 dk = \frac{\Omega}{(2\pi)^3}4\pi k.kdk \tag{6.62}$$

$$= \frac{\Omega}{2\pi^2}\sqrt{\frac{2m\varepsilon}{\hbar^2}}\frac{m}{\hbar^2}d\varepsilon \tag{6.63}$$

Thus a density of states in energy is obtained, defined by $dn(\varepsilon) = D(\varepsilon)\,d\varepsilon$, with

$$D(\varepsilon) = C\Omega\sqrt{\varepsilon} \text{ with } C = \frac{(2s+1)}{4\pi^2}\left(\frac{2m}{\hbar^2}\right)^{3/2} \tag{6.64}$$

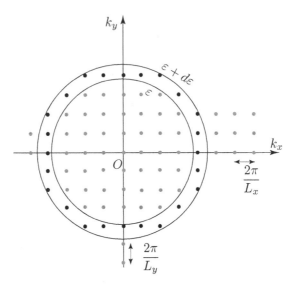

FIG. 6.5: Number of states of energy between ε and $\varepsilon + d\varepsilon$, (two-dimensional problem).

This expression of the energy density of states, which includes the spin degeneracy, is valid for any positive energy; the density of states vanishes for $\varepsilon < 0$, since then no state can exist, the energy being only of kinetic origin for the considered free particle.

c) Densities of States (Stationary-Waves Conditions)

We have seen that, in the case of stationary-wave conditions, all the physically distinct solutions are obtained when restricting the wave vectors choice to those with all three positive components. The volume corresponding to the states of energies between ε and $\varepsilon + d\varepsilon$ must thus be limited to the corresponding trihedral of the $\vec{k}$-space. This is the 1/8th of the space between the two spheres of radii k and $k + dk$, i.e., $\frac{1}{8} 4\pi k^2 dk$, dk being still related to $d\varepsilon$ by

$$d\varepsilon = \frac{\hbar^2}{m} k dk \qquad (6.65)$$

However the wave vector density of states is eight times larger for the stationary-wave quantization condition than for the B-VK condition (the allowed points of the $\vec{k}$-space are twice closer in the x, y and z directions). Then one obtains a number of orbital states, without including the spin degeneracy, of

$$dn' = \frac{\Omega}{\pi^3} \frac{1}{8} 4\pi k^2 dk \qquad (6.66)$$

The density $D(\varepsilon)$ *thus has exactly the same value than in* b) when the spin degeneracy $(2s + 1)$ is introduced.

One sees that both types of quantization condition are equivalent for a macroscopic system. In most of the remaining parts of this course, the B-VK conditions will be used, as they are simpler because they consider the whole wave vector space. Obviously, in specific physical conditions, one may have to use the stationary-wave quantization conditions.

Notes :

In presence of a magnetic potential energy, which does not affect the orbital part of the wave function, the density of states $D_{\vec{k}}(\vec{k})$ is unchanged. On the contrary, $D(\varepsilon)$ is shifted (see § 7.1.3).

As an exercise, we now calculate the energy density of states for *free particles* in the two- or one-dimension space. The relations between ε and k, between $d\varepsilon$ and dk are not modified [(6.61) and (6.65)].

  – The two-dimension elementary area is $2\pi k dk$, the $\vec{k}$ -density of states is given by $\dfrac{L_x L_y}{(2\pi)^2}$ without taking the spin in account and the energy density of states is a constant.

  – In one dimension, the elementary length is $2dk_x$ (k_x and $-k_x$ correspond to the same energy, whence the factor 2), the k_x-density of states is $\dfrac{L}{2\pi}$ without spin, and $D(\varepsilon)$ varies like $\dfrac{1}{\sqrt{\varepsilon}}$. $D(\varepsilon)$ diverges for ε tending to zero but its integral, which gives the total number of accessible states from $\varepsilon = 0$ to $\varepsilon = \varepsilon_0$, is finite.

6.5   Fermi-Dirac Distribution ; Bose-Einstein Distribution

We have just seen that, for a macroscopic system, the accessible energy levels are extremely close and have defined a density of states, which is a discrete function of energy ε : in the large volume limit, it is as if $D(\varepsilon)$ was a function of the continuous variable ε.

Let us now return to Statistical Physics in the framework of the large volumes limit. From the average number of occupation $\langle n_k \rangle$ (see § 6.3) of an energy level ε_k, one defines the energy function " occupation factor" or "distribution," of Fermi-Dirac or of Bose-Einstein according to the nature of the considered particles :

$$f_{FD}(\varepsilon) = \frac{1}{e^{\beta(\varepsilon-\mu)} + 1} = \frac{1}{e^{\beta\varepsilon-\alpha} + 1} \qquad (6.67)$$

$$f_{BE}(\varepsilon) = \frac{1}{e^{\beta(\varepsilon-\mu)} - 1} = \frac{1}{e^{\beta\varepsilon-\alpha} - 1} \qquad (6.68)$$

Recall that α and the chemical potential μ are related by $\alpha = \beta\mu$ with $\beta = \frac{1}{k_B T}$.

The consequences of the expression of their distribution will be studied in details in chapter 7 for free fermions and in chapter 9 for bosons.

6.6   Average Values of Physical Parameters at T in the Large Volumes Limit

On the one hand, the energy density of states $D(\varepsilon)$ summarizes the Quantum Mechanical properties of the system, i.e., it expresses the solutions of the Schroedinger equation for an individual particle. On the other hand, the distribution $f(\varepsilon)$ expresses the statistical properties of the considered particles. To calculate the average value of a given physical parameter, one writes that in the energy interval between ε and $\varepsilon + d\varepsilon$, the *number of accessible states* is $D(\varepsilon)d\varepsilon$, and that at temperature T and for a parameter α or a chemical potential μ, these levels are *occupied* according to the *distribution* $f(\varepsilon)$ (of Fermi-Dirac or Bose-Einstein). Thus

$$N = \int_{-\infty}^{+\infty} D(\varepsilon)f(\varepsilon)\, d\varepsilon \qquad (6.69)$$

$$U = \int_{-\infty}^{+\infty} \varepsilon D(\varepsilon)f(\varepsilon)\, d\varepsilon \qquad (6.70)$$

The temperature and the chemical potential are included inside the distribution expression. In fact the first equation allows one to determine the chemical potential value.

To obtain the grand potential, one generalizes to a large system relations (6.31) and (6.32) that were established for discrete parameters :

$$A = +k_B T \int_{-\infty}^{+\infty} D(\varepsilon)\ln(1 - f(\varepsilon))\, d\varepsilon \text{ for fermions} \qquad (6.71)$$

$$A = -k_B T \int_{-\infty}^{+\infty} D(\varepsilon)\ln(1 + f(\varepsilon))\, d\varepsilon \text{ for bosons} \qquad (6.72)$$

6.7   Common Limit of the Quantum Statistics

6.7.1   Chemical Potential of the Ideal Gas

In the case of free *particles* (fermions or bosons) of spin s, *confined* in a volume Ω, one expresses the number of particles (6.69) using the expression (6.64) of the density of states $D(\varepsilon)$:

$$N = \frac{(2s+1)\Omega(2m)^{3/2}}{4\pi^2\hbar^3} \int_0^{+\infty} \frac{\sqrt{\varepsilon}\,d\varepsilon}{e^{\beta\varepsilon-\alpha} \mp 1} \tag{6.73}$$

where, in the denominator, the sign $-$ is for bosons and the sign $+$ for fermions. If the temperature is finite, the dimensionless variable of the integral is $\beta\varepsilon = x$, so that this relation can be written :

$$\frac{N}{\Omega} = \left(\frac{2mk_BT}{\hbar^2}\right)^{3/2} \frac{(2s+1)}{4\pi^2} \int_0^{+\infty} \frac{\sqrt{x}\,dx}{e^{x-\alpha} \mp 1} \tag{6.74}$$

The physical parameters are then put into the left member, which depends of the system density and temperature as

$$\frac{N}{\Omega}\frac{4\pi^2}{(2\pi)^3}\frac{1}{2s+1}\frac{1}{2^{3/2}}\left(\frac{h^2}{mk_BT}\right)^{3/2} = \frac{N}{\Omega}\frac{\sqrt{\pi}}{2}\frac{1}{2s+1}(\lambda_{\text{th}})^3 = \text{constant}\ \frac{N}{\Omega}\lambda_{\text{th}}^3 \tag{6.75}$$

The thermal de Broglie wavelength λ_{th} appears, as defined in §4.3. The quantity (6.75) must be equal to the right member integral, which is a function of α such that

$$u_\mp(\alpha) = \int_0^\infty \frac{\sqrt{x}\,dx}{e^{x-\alpha}\mp 1} \tag{6.76}$$

In fact (6.75) expresses, to the factor $\sqrt{\pi}/2(2s+1)$, the cube of the ratio of the thermal de Broglie wavelength λ_{th} to the average distance between particles $d = (\Omega/N)^{1/3}$. It is this comparison which, in §4.3, had allowed us to define the limit of the classical ideal gas model. Since (6.75) and (6.76) are equal, one obtains :

$$u_\mp(\alpha) = \frac{1}{2s+1}\frac{\sqrt{\pi}}{2}\frac{N}{\Omega}(\lambda_{\text{th}})^3 \tag{6.77}$$

The integral $u_\mp(\alpha)$ can be numerically calculated versus α, it is a monotonous increasing function of α for either Fermi-Dirac (FD) or Bose-Einstein (BE) Quantum Statistics (Fig. 6.6). We already know the sign or the value of $u(\alpha)$ for some specific physical situations :

It was noted in §6.2.2 that the expression of $f(\varepsilon)$ for Bose-Einstein statistics implies that $\alpha < \beta\varepsilon_1$, i.e., $\alpha < 0$ for free bosons, since then ε_1 is zero. We

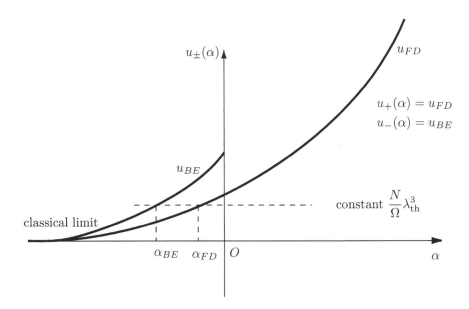

FIG. 6.6: Sketch of the function $u_\pm(\alpha)$ and determination of α from the physical data of the considered system.

will see in chapter 9 of this book that, in some situations, α vanishes, thus producing the *Bose-Einstein condensation*, which is a phase transition.

In the case of *free fermions at low temperature*, it will be seen in the next chapter that μ is larger than ε_1, which is equal to zero, i.e., $\alpha > 0$.

When α tends to minus infinity, the denominator term takes the *same expression* for both *fermions* and *bosons*. The integration is exactly performed :

$$u_\mp(\alpha) \simeq e^\alpha \int_0^{+\infty} \sqrt{x}\,e^{-x}dx = e^\alpha \frac{\sqrt{\pi}}{2} \ll 1 \qquad (6.78)$$

Considering the expression of the right member of 6.76, this limit does occur if the particle density $n = N/\Omega$ tends to zero or the temperature tends to infinity.

In the framework of this classical limit, one deduces

$$e^\alpha = \frac{N}{\Omega}\frac{\lambda_{th}^3}{(2s+1)} \qquad (6.79)$$

which allows one to express the chemical potential μ, since $\alpha = \mu/k_B T$. One finds, if the spin degeneracy is neglected, i.e., if one takes $2s + 1 = 1$,

$$\mu = -k_B T \ln \left[\frac{\Omega}{N} \left(\frac{2\pi m k_B T}{h^2} \right)^{3/2} \right] \tag{6.80}$$

the very expression obtained in § 4.4.3 for the ideal gas.

6.7.2 Grand Canonical Partition Function of the Ideal Gas

Let us look for the expression of A in the limit where α tends to minus infinity. For both fermions and bosons, one gets the same limit, by developing the Naperian logarithm to first order : it is equal to

$$A = -k_B T \sum_k e^{\alpha - \beta \varepsilon_k} \tag{6.81}$$

$$= -k_B T e^\alpha \sum_k e^{-\beta \varepsilon_k} \tag{6.82}$$

One recognizes the one-particle partition function

$$z_1 = \sum_k e^{-\beta \varepsilon_k} \tag{6.83}$$

(its value for the ideal gas was calculated in § 4.4.1), whence

$$A = -k_B T e^\alpha z_1 \tag{6.84}$$

By definition

$$A = -k_B T \ln Z_G, \tag{6.85}$$

thus in the common limit of both Quantum Statistics, i.e., in the *classical limit* conditions, Z_G is equal to

$$Z_G = \exp(e^\alpha z_1) = \sum_N e^{\alpha N} \frac{z_1^N}{N!} \tag{6.86}$$

Now the general definition of the grand canonical partition function Z_G, using the N-particle canonical partition function Z_{cN}, is

$$Z_G = \sum_N e^{\alpha N} Z_{cN} \tag{6.87}$$

This shows that in the case of *identical* particles, for sake of consistency, the N-particles canonical partition function must contain the factor $1/N!$, the "memory" of the Quantum Statistics, to take into account the indistinguishability of the particles :

$$Z_{cN} = \frac{1}{N!} z_1^N \tag{6.88}$$

This factor ensures the extensivity of the thermodynamical functions, like S, and allows to solve the Gibbs paradox (there is no extra entropy when two volumes of gases of the *same density and the same nature* are allowed to mix into the same container.) (see § 4.4.3).

Notes :

The above reasoning allows one to grasp how, in the special case of free particles, one goes from a quantum regime to a classical one.

We did not specify to which situation corresponds the room temperature, the most frequent experimental condition. We will see that for metals, for example, it corresponds to a very low temperature regime, of quantum nature (see chapter 7).

Summary of Chapter 6

To account for the occupation conditions of the energy states as stated by the Pauli principle, it is simpler to work in the grand canonical ensemble, in which the number of particles is only given in average : then, at the end of the calculation, the value of the chemical potential is adjusted so that this average number coincides with the real number of particles in the system. As soon as the system is macroscopic, the fluctuations around the average particle number predicted in the grand canonical ensemble are much smaller than the resolution of any measure.

The grand partition function gets factorized by energy level and takes a different expression for fermions and for bosons.

Remember the *distribution*, which provides the average number of particles, of chemical potential equal to μ, on the energy level ε at temperature $T = 1/k_B\beta$:

for fermions

$$f_{FD}(\varepsilon) = \frac{1}{\exp\beta(\varepsilon - \mu) + 1} \tag{6.89}$$

for bosons

$$f_{BE}(\varepsilon) = \frac{1}{\exp\beta(\varepsilon - \mu) - 1} \tag{6.90}$$

For a free particle in a box, we calculated the location of the allowed energy levels. The periodic boundary conditions of Born-Von Kármán (or in an analogous way the stationary-waves conditions) evidence quantized levels, and for a box of macroscopic dimensions ("large volume limit"), the allowed levels are extremely close. One then defines an energy density of states such that the number of allowed states between ε and $\varepsilon + d\varepsilon$ is equal to

$$dn = D(\varepsilon)d\varepsilon \tag{6.91}$$

For a free particle in the three-dimensional space $D(\varepsilon) \propto \sqrt{\varepsilon}$.

In the large volume limit the average values at temperature T of physical parameters are expressed as integrals, which account for the existence of allowed levels through the energy density of states and for their occupation by the particles at this temperature through the distribution term :

$$N = \int_{-\infty}^{+\infty} D(\varepsilon)f(\varepsilon)\,d\varepsilon \qquad\qquad (6.92)$$

$$U = \int_{-\infty}^{+\infty} \varepsilon D(\varepsilon)f(\varepsilon)\,d\varepsilon \qquad\qquad (6.93)$$

We showed how for free particles the Classical Statistics of the ideal gas is the common limit of the Fermi-Dirac and Bose-Einstein statistics when the temperature is very high and/or the density very low. The factor $1/N!$ in the canonical partition function of the ideal gas is then the "memory" of the Quantum Statistics properties.

Chapter 7

Free Fermions Properties

We now come to the detailed study of the properties of the Fermi-Dirac quantum statistics. First at zero temperature we will study the behavior of free fermions, which shows the consequences of the Pauli principle. Then we will be concerned by electrons, considered as free, and will analyze two of their properties which are temperature-dependent, i.e., their specific heat and the " thermionic" emission. In the next chapter we will understand why this free electron model indeed properly describes the properties of the conduction electrons in metals.

In fact, in the present chapter we follow the historical order : the properties of electrons in metals, described by this model, were published by Arnold Sommerfeld (electronic specific heat, Spring 1927) and Wolfang Pauli (electrons paramagnetism, December 1926), shortly after the statement of the Pauli principle at the beginning of 1925.

7.1 Properties of Fermions at Zero Temperature

7.1.1 Fermi Distribution, Fermi Energy

The Fermi-Dirac distribution, presented in § 6.5, gives the average value of the occupation number of the one-particle state of energy ε, and is a function of the chemical potential μ. Its expression is

$$f(\varepsilon) = \frac{1}{e^{\beta(\varepsilon-\mu)} + 1} = \frac{1}{2}\left[1 - \tanh\frac{\beta(\varepsilon-\mu)}{2}\right] \qquad (7.1)$$

It is plotted in Fig. 7.1 and takes the value $\dfrac{1}{2}$ for $\varepsilon = \mu$ for any value of β and thus of the temperature. This particular point is a center of inversion of the curve, so that the probability of "nonoccupation" for $\mu - \delta\varepsilon$, i.e., $1 - f(\mu - \delta\varepsilon)$, has the same value as the occupation probability for $\mu + \delta\varepsilon$, i.e., $f(\mu + \delta\varepsilon)$. At the point $\varepsilon = \mu$ the tangent slope is $-\dfrac{\beta}{4}$: the lower the temperature, the larger this slope.

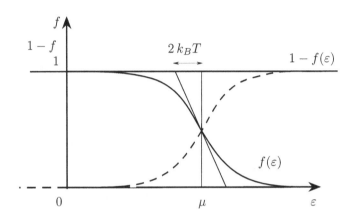

FIG. 7.1 : Variation versus energy of $f(\varepsilon)$ and $1 - f(\varepsilon)$.

In particular when T tends to zero, $f(\varepsilon)$ tends to a "step function" (Heaviside function). One usually calls *Fermi energy* the *limit* of μ when T *tends to zero* ; it is noted ε_F, its existence and its value will be specified below. Then one obtains at $T = 0$:

$$\begin{cases} f(\varepsilon) = 1 & \text{if } \varepsilon < \varepsilon_F \\ f(\varepsilon) = 0 & \text{if } \varepsilon > \varepsilon_F \end{cases} \tag{7.2}$$

At $T = 0$ K the filled states thus begin at the lowest energies, and end at ε_F, the last occupied state. The states above ε_F are empty. This is in agreement with the Pauli principle : indeed in the fundamental state the particles are in their state of minimum energy but two fermions from the same physical system cannot occupy the same quantum state, whence the requirement to climb up in energy until the particles all located.

In the framework of the large volume limit, to which this course is limited, the total number N of particles in the system satisfies

$$N = \int_0^\infty D(\varepsilon) f(\varepsilon) d\varepsilon \tag{7.3}$$

where $D(\varepsilon)$ in the one-particle density of states in energy. This condition determines the μ value.

At zero temperature, the chemical potential, then called the "Fermi energy ε_F,", is thus deduced from the condition that

$$N = \int_0^{\varepsilon_F} D(\varepsilon)d\varepsilon \tag{7.4}$$

The number N corresponds to the hatched area of Fig. 7.2.

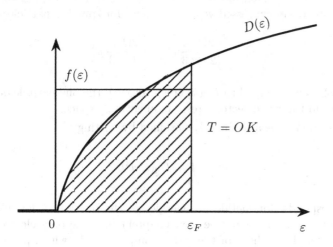

FIG. 7.2: Density of states $D(\varepsilon)$ of three-dimensional free particles and determination of the Fermi energy at zero temperature. The hatched area under the curve provides the number N of particles in the system.

For a three-dimensional system of free particles, with a spin $s = 1/2$, from (6.64)

$$D(\varepsilon) = \frac{\Omega}{2\pi^2\hbar^3}(2m)^{3/2}\sqrt{\varepsilon} \quad \text{for} \quad \varepsilon > 0$$
$$D(\varepsilon) = 0 \qquad\qquad\qquad \text{for} \quad \varepsilon \leq 0 \tag{7.5}$$

The integral (7.4) is then equal to

$$N = \frac{2}{3}\frac{\Omega}{2\pi^2\hbar^3}(2m)^{3/2}\varepsilon_F^{3/2} \tag{7.6}$$

that is,

$$\varepsilon_F = \frac{\hbar^2}{2m}\left(3\pi^2\frac{N}{\Omega}\right)^{2/3} \tag{7.7}$$

This expression leads to several comments :

- N and Ω are extensive, thus ε_F is intensive;
- ε_F only depends on the density N/Ω;
- the factor of $\hbar^2/2m$ in (7.7) is equal to k_F^2, where k_F is the Fermi wave vector : in the $\vec{k}$-space the surface corresponding to the constant energy ε_F, called "Fermi surface," contains all the filled states. In the case of free particles, it is a sphere, the "Fermi sphere," of radius k_F. One can also define the momentum $p_F = \hbar k_F = \sqrt{2m\varepsilon_F}$. Then expression (7.6) can be obviously expressed versus k_F or p_F for spin-1/2 particles :

$$N = \frac{2\Omega}{(2\pi)^3}\frac{4}{3}\pi k_F^3 = \frac{2\Omega}{h^3}\frac{4\pi}{3}p_F^3 \tag{7.8}$$

The first factor is equal to $D_{\vec{k}}(\vec{k})$ (resp. $D_{\vec{p}}(\vec{p})$), the one-particle density of states in the wave vector- (resp. momentum-) space.

Expression (7.8) is also directly obtained by writing :

$$N = \int_0^{k_F} D_{\vec{k}}(\vec{k})d^3\vec{k} \ .$$

- ε_F corresponds, for metals, to energies of the order of a few eV. Indeed, let us assume now (see next chapter) that the only electrons to be considered are those of the last incomplete shell, which participate in the conduction. Take the example of copper, of atomic mass 63.5 g and density 9 g/cm³. There are thus $9\mathcal{N}/63.5 = 0.14\mathcal{N}$ copper atoms per cm³, $\mathcal{N}$ being the Avogadro number. Since one electron per copper atom participates in conductivity, the number of such electrons per cm³ is of $0.14\mathcal{N} = 8.4 \times 10^{22}/\text{cm}^3$ i.e., $8.4 \times 10^{28}/ \text{m}^3$. This corresponds to a wave vector $k_F = 1.4 \times 10^{10}$ m^{-1}, and to a velocity at the Fermi level $v_F = \dfrac{\hbar k_F}{m}$ of 1.5×10^6 m/sec. One deduces $\varepsilon_F = 7$ eV.

Remember that the thermal energy at 300 K, $k_B T$, is close to 25 meV, so that the Fermi energy ε_F corresponds to a temperature T_F such that $k_B T_F = \varepsilon_F$, in the 10^4 or 10^5 K range according to the metal (for copper 80, 000 K)!

7.1.2 Internal Energy and Pressure of a Fermi Gas at Zero Temperature

The internal energy at zero temperature is given by

$$U = \int_0^{\varepsilon_F} \varepsilon D(\varepsilon)d\varepsilon \tag{7.9}$$

For the three-dimensional density of states (7.5), U is equal to :

$$U = \frac{2}{5} \frac{\Omega}{2\pi^2 \hbar^3} (2m)^{3/2} \varepsilon_F^{5/2} \tag{7.10}$$

i.e., by expressing N using (7.6),

$$U = \frac{3}{5} N \varepsilon_F \tag{7.11}$$

Besides,

$$P = -\left(\frac{\partial U}{\partial \Omega}\right)_N \tag{7.12}$$

From (7.10) and (7.7), U varies like $N^{5/3}\Omega^{-2/3}$, so that at constant particles number N

$$P = \frac{2}{3}\frac{U}{\Omega} \tag{7.13}$$

For copper, one obtains $P = 3.8 \times 10^{10} \text{ N/m}^2 = 3.7 \times 10^5$ atmospheres. This very large pressure, exerted by the particles on the walls of the solid which contains them, exists even at zero temperature. Although the state equation (7.13) is identical in the cases of fermions and of the ideal gas, the two situations are very different : for a classical gas, when the temperature tends to zero, both the internal energy and the pressure vanish. For fermions, the Pauli principle dictates that levels of nonzero energies are filled. The fact that electrons in the same orbital state cannot be more than a pair, and of opposite spins, leads to their mutual repulsion and to a large average momentum, at the origin of this high pressure.

7.1.3 Magnetic Properties. Pauli Paramagnetism

A gas of electrons, each of them carrying a spin magnetic moment equal to the Bohr magneton $\vec{\mu}_B$, is submitted to an external magnetic field $\vec{B}$. If N_+ is the number of electrons with a moment projection along $\vec{B}$ equal to $+\mu_B$, N_- that of moment $-\mu_B$, the gas total magnetic moment (or magnetization) along $\vec{B}$ is

$$\mathcal{M} = \mu_B(N_+ - N_-) \tag{7.14}$$

The density of states $D_{\vec{k}}(\vec{k})$ is the same as for free electrons. Since the potential energy of an electron depends on the direction of its magnetic moment, the total energy of an electron is

$$\varepsilon_\pm = \frac{\hbar^2 \vec{k}^2}{2m} \mp \mu_B B \tag{7.15}$$

The densities of states $D_+(\varepsilon)$ and $D_-(\varepsilon)$ now differ for the two magnetic moment orientations, whereas in the absence of magnetic field, $D_+(\varepsilon) = D_-(\varepsilon) = D(\varepsilon)/2$: here

$$\begin{cases} D_+(\varepsilon) & = \frac{1}{2}D(\varepsilon + \mu_B B) \\ D_-(\varepsilon) & = \frac{1}{2}D(\varepsilon - \mu_B B) \end{cases} \tag{7.16}$$

Besides, the populations of electrons with the two orientations of magnetic moment are in equilibrium, thus they have the same chemical potential.

The solution of this problem is a standard exercise of Statistical Physics. Here we just give the result, in order to comment upon it : one obtains the magnetization

$$\mathcal{M} = \mu_B \cdot \mu_B B \cdot D(\varepsilon_F) \tag{7.17}$$

This expresses that the moments with the opposite orientations compensate, except for a small energy range of width $\mu_B B$ around the Fermi energy. The corresponding (Pauli) susceptibility, defined by $\lim\limits_{\vec{B}\to 0} \dfrac{1}{\mu_0 \Omega} \dfrac{\vec{M}}{\vec{B}}$, is equal to

$$\chi = \frac{1}{\mu_0} \frac{3}{2} \frac{N}{\Omega} \frac{\mu_B^2}{\varepsilon_F} = \frac{1}{\mu_0} \frac{3}{2} \frac{N}{\Omega} \frac{\mu_B^2}{k_B T} \cdot \frac{k_B T}{\varepsilon_F} \tag{7.18}$$

with $\mu_0 = 4\pi \times 10^{-7}$. It is reduced with respect to the classical result (Curie law) by a factor of the order of $k_B T/\varepsilon_F$, which expresses that only a very small fraction of the electrons participate in the effect (in other words, only a few electrons can reverse their magnetic moment to have it directed along the external field).

7.2   Properties of Fermions at Non-Zero Temperature

7.2.1   Temperature Ranges and Chemical Potential Variation

At nonvanishing temperature, one again obtains the chemical potential μ position from expression (7.3), which links μ to the total number of particles :

$$\begin{aligned} N &= \int_0^\infty D(\varepsilon) f(\varepsilon) d\varepsilon \\ &= \int_0^\infty D(\varepsilon) \frac{1}{e^{\beta(\varepsilon-\mu)} + 1} d\varepsilon \end{aligned} \tag{7.3}$$

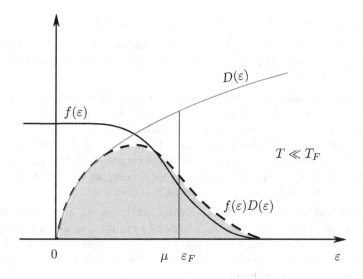

FIG. 7.3 : Determination of the chemical potential μ for $T \ll T_F$.

Relation (7.3) implies that μ varies with T. Fig. 7.3 qualitatively shows for $k_B T \ll \varepsilon_F$ the properties which will be calculated in Appendix 7.1. Indeed the area under the curve representing the product $D(\varepsilon)f(\varepsilon)$ gives the number of particles. At $T = 0$ K, one has to consider the area under $D(\varepsilon)$ up to the energy ε_F ; at small T, a few states above μ are filled and an equivalent number below μ are empty. Since the function $D(\varepsilon)$ is increasing, to maintain an area equal to N, μ has to be smaller than ε_F, of an amount proportional to $D'(\mu)$.

The exact relation allowing one to deduce $\mu(T)$ for the two Quantum Statistics was given in (6.74) : recalling that $\alpha = \beta\mu$, and substituting (7.5) into (7.3) one obtains for fermions

$$\int_0^\infty \frac{\sqrt{x}\,dx}{\exp(x - \alpha) + 1} = 2\pi^2 \cdot \frac{N}{\Omega} \left(\frac{\hbar^2}{2mk_BT} \right)^{3/2} \tag{7.19}$$

$$= \frac{2}{3} \left(\frac{\varepsilon_F}{k_BT} \right)^{3/2} \tag{7.20}$$

In the integral we defined $\beta\varepsilon = x$.

Thus α is related to the quantity $\varepsilon_F/k_BT = T_F/T$.

  – When T vanishes, the integral (7.19) tends to infinity : the exponential in the denominator no longer plays any role since α tends to infinity (ε_F is finite and β tends to infinity). The right member also tends to infinity. One thus has to return to equation (7.4).

– We have just studied the low temperature regime $T \ll T_F$, which takes place in metals at room temperature ($T_F \sim 10^4$ K) and corresponds to Fig. 7.3.

  In astrophysics, the "white dwarfs" enter in this framework. They are stars at the end of their existence, which have already burnt all their nuclear fuel and therefore have shrunk. Their size is of the order of a planet (radius of about 5000 km) for a mass of 2×10^{30} kg (of the order of that of the sun), their temperature T of 10^7 K. Their high density in electrons determines ε_F. T_F being in the 10^9 K range, the low temperature regime is achieved!

  This regime for which the Quantum Statistics apply is called the "degenerate" (in opposition to "classical") regime : as already mentioned about Pauli paramagnetism, and will be seen again below, the "active" particles all have an energy close to μ, this is almost as if this level was degenerate in the Quantum Mechanics meaning.

– On the other hand, when $T \gg T_F$, the integral (7.19) vanishes. This is realized either at low density (ε_F small), or at given density when T tends to infinity. Then $e^{-\alpha}$ tends to infinity, and since $\alpha = \beta\mu$ with β tending to zero, this implies that μ tends to $-\infty$. One then finds the Maxwell-Boltzmann Classical Statistics, as discussed in § 6.7.2.

– Between these two extreme situations, μ thus decreases from ε_F to $-\infty$ when the ratio T/T_F increases.

When studying systems at room temperature, the regime is degenerate in most cases. For properties mainly depending on the Pauli principle, one can consider then to be at zero temperature. This is the situation when *filling the electronic shells* in *atoms* according to the "Aufbau (construction) principle", widely used in Chemistry courses : the Quantum Mechanics problem is first solved in order to find the one-electron energy levels of the considered atom. The solutions ranked in increasing energies are the states $1s, 2s, 2p, 3s, 3p, 4s, 3d$, etc. (except some particular cases for some transition metals). These levels are then filled with electrons, shell after shell, taking into account the level degeneracy (K shell, $n = 1$, containing at most two electrons, L shell of $n = 2$, with at most eight electrons, etc.). Since the energy splittings between levels are of the order of a keV for the deep shells and of an eV for the valence shell, the thermal energy remains negligible with respect to the energies coming into play and it is as if the temperature were zero.

On the other hand, if one is interested in temperature-dependent properties, like the specific heat or the entropy, which are derivatives versus temperature of thermodynamical functions, more exact expressions are required and at low temperature one will proceed through limited developments versus the variable T/T_F.

7.2.2   Specific Heat of Fermions

The very small contribution of electrons to a metal specific heat, which follow
the Dulong and Petit law (see Ch. 2, § (2.4.5)) only introducing the vibrations
of the lattice atoms, presented a theoretical challenge in the 1910s. This lead
to a questioning of the electron conduction model, as proposed by Paul Drude
in 1900.

To calculate the specific heat $C_v = \left(\dfrac{\partial U}{\partial T}\right)_{N,\Omega}$ of a fermions assembly, one
needs to express the temperature dependence of the internal energy.

One can already make an estimate of C_v from the observation of the shape of
the Fermi distribution : the only electrons which can absorb energy are those
occupying the states in a region of a few $k_B T$ width around the chemical
potential, and each of them will increase its energy by an amount of the order
$k_B T$ to reach an empty state.

The number of concerned electrons is thus $\sim k_B T \cdot D(\mu)$ and for a three-
dimensional system at low temperature ($\dfrac{k_B T}{\varepsilon_F} \ll 1$, so that μ and ε_F are
close),

$$N \sim \int_0^\mu D(\varepsilon)d\varepsilon \propto \mu^{3/2} , \qquad (7.21)$$

$$\text{i.e.,} \quad \frac{dN}{d\mu} = \frac{3}{2}\frac{N}{\mu} = D(\mu) \qquad (7.22)$$

The total absorbed energy is thus of the order of

$$U \sim \left(k_B T \cdot \frac{3}{2}\frac{N}{\mu}\right) \cdot k_B T \qquad (7.23)$$

Consequently,

$$C_v = \frac{dU}{dT} \sim N k_B \left(\frac{k_B T}{\mu}\right) \qquad (7.24)$$

These considerations evidence a reduction of the value of C_v with respect to
the classical value $N k_B$ in a ratio of the order $k_B T/\mu \ll 1$, because of the
very small number of concerned electrons.

The exact approach refers to a general method of calculation of the integrals in which the Fermi distribution enters : the Sommerfeld development.
The complete calculation is given in Appendix 7.1. Here we just indicate its
principle.

At a temperature T small with respect to T_F, one looks for a limited development in T/T_F of the fermions internal energy U, which depends of T, and of

the chemical potential, itself temperature-dependent. The searched quantity $C_v = \left(\dfrac{dU}{dT}\right)_{N,\Omega}$ is then equal to

$$C_v = \frac{3}{2}Nk_B \cdot \frac{\pi^2}{3}\left(\frac{k_BT}{\varepsilon_F}\right) \tag{7.25}$$

which is indeed of the form (7.24), as far as μ is not very different from ε_F (see Appendix 7.1).

The first factor in (7.25) is the specific heat of a monoatomic ideal gas; the last factor is much smaller than unity in the development conditions, it ranges between 10^{-3} and 10^{-2} at 300 K (for copper it is 3.6×10^{-3}) and expresses the small number of effective electrons.

The electron's specific heat in metals is measured in experiments performed at temperatures in the kelvin range. In fact the measured total specific heat includes the contributions of both the electrons and the lattice

$$C_v^{\text{total}} = C_v^{\text{el}} + C_v^{\text{lattice}} \tag{7.26}$$

The contribution C_v^{lattice} of the lattice vibrations to the solid specific heat was studied in Ch. 2, § 2.4.5. The Debye model was mentioned, which correctly interprets the low temperature results by predicting that C_v^{lattice} is proportional to T^3, as observed in non-metallic materials. The lower the temperature, the less this contribution will overcome that of the electrons.

The predicted variation versus T for C_v^{total} is thus given by

$$C_v^{\text{total}} = \gamma T + aT^3 \tag{7.27}$$

$$\frac{C_v^{\text{total}}}{T} = \gamma + aT^2 \tag{7.28}$$

with, from (7.25),

$$\gamma = Nk_B\frac{\pi^2}{2}\frac{k_B}{\varepsilon_F} \tag{7.29}$$

From the measurement of γ one deduces a value of ε_F of 5.3 eV for copper whereas in § 7.1.1 we calculated 7 eV from the value of the electrons concentration, using (7.7). In the case of silver the agreement is much better since the experiment gives $\varepsilon_F = 5.8$ eV, close to the prediction of (7.7).

The possible discrepancies arise from the fact that the mass m used in the theoretical expression of the Fermi energy was that of the free electron. Now a conduction electron is submitted to interactions with the lattice ions, with the solid vibrations and the other electrons. Consequently, its motion is altered, which can be expressed by attributing it an *effective mass* m^*. This concept will be explained in §8.2.3 and §8.3.3.

Defining

$$\varepsilon_F = \frac{\hbar^2}{2m^*} \left(3\pi^2 \frac{N}{V}\right)^{2/3}, \tag{7.30}$$

one deduces from these experiments a mass $m^* = 1.32m$ for copper, $m^* = 1.1m$ for silver, where m is the free electron mass.

Note that the fermion entropy can be easily deduced from the development of the Appendix : we saw that $A = -P\Omega$ (see §3.5.3), so that, according to (7.13), $A = -\frac{2}{3}U$. Besides, from (3.5.3), $S = -\left(\frac{\partial A}{\partial T}\right)_{\mu,\Omega}$. One deduces

$$S = Nk_B \frac{\pi^2}{2} \frac{k_B T}{\varepsilon_F} = C_v \tag{7.31}$$

The electrons entropy thus satisfies the third law of Thermodynamics since it vanishes with T.

7.2.3 Thermionic Emission

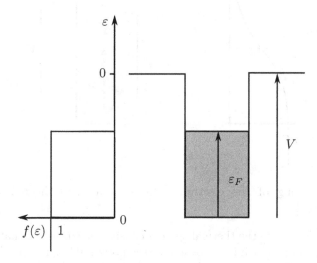

FIG. 7.4: Potential well for the electrons of a metal. In order to extract an electron at zero temperature, one illuminates the solid with a photon of energy higher than $V - \varepsilon_F$.

The walls of the "potential box" which confines the electrons in a solid are not really infinitely high. The photoelectrical effect, evidenced by Heinrich

Hertz (1887), and interpreted by Albert Einstein (this was the argument for Einstein's Nobel prize in 1921), allows one to extract electrons from a solid, at room temperature, by illumination with a light of high enough energy : for zinc, a radiation in the near ultraviolet is required (Fig. 7.4).

The energy's origin is here the state of a free electron with zero kinetic energy, i.e., with such a convention the Fermi level has a negative energy. The potential barrier is equal to V. To extract an electron at *zero temperature*, a photon must bring it at least $V - \varepsilon_F$, an energy usually in the ultraviolet range.

When heating this metal to a *high temperature* T, the Fermi distribution is spreading toward little negative or even positive energies, so that a few electrons are then able to pass the potential barrier and become free (Fig. 7.5). One utilizes tungsten heated to about 3000 K to make the filaments of the electrons cannons of TV tubes or electronic microscopes, as its fusion temperature is particularly high (3700 K).

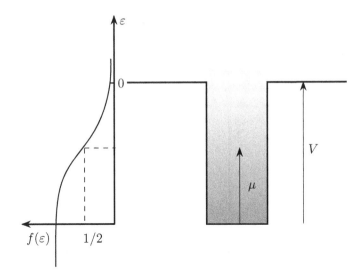

FIG. 7.5: Filling of the electronic states in a metal heated to a high temperature T.

One way to calculate the thermal current consists in expressing that a particle escapes the metal only if it can overcome the potential barrier, i.e. if its kinetic energy in the direction perpendicular to the surface, taken as z axis, is large enough. It should satisfy

$$\frac{p^2}{2m} \geq \frac{p_z^2}{2m} \geq V \tag{7.32}$$

We know that the electronic energies, and thus V and $V - \mu$, are of the order of a few eV : whatever the filament temperature, the system is in a regime where $k_B T$ remains small with respect to the concerned energies, and where $\frac{k_B T}{\varepsilon_F} \ll 1$. Consequently, μ is still close to ε_F, and in the Fermi distribution

$$f(\varepsilon) = \frac{1}{\exp \beta \left(\frac{p^2}{2m} - \mu \right) + 1} \tag{7.33}$$

the denominator exponent is large for the electrons likely to be emitted.

The Fermi distribution can then be approximated by a Maxwell-Boltzmann distribution

$$f(\varepsilon) \simeq e^{\beta \mu} e^{-\beta \frac{p^2}{2m}} \tag{7.34}$$

There remains to estimate the number Δn of electrons moving toward the surface ΔS during a time Δt. This is a standard problem of kinetic theory of gases (the detailed method of calculation is analogous to the kinetic calculation of the pressure, § 4.2.2).

One obtains a total current $j_z = \Delta n / \Delta S \Delta t$ equal to

$$j_z = \frac{\Delta n}{\Delta S \Delta t} = \frac{2}{h^3} 2\pi m (k_B T)^2 e^{-\beta(V - \mu)} \tag{7.35}$$

This expression shows the very fast variation of the thermal emission current with the temperature T, in $T^2 e^{-(V-\mu)/k_B T}$. It predicts that a 0.3×0.3 mm^2 tungsten tip emits a current of 15 mA at 3000 K and of only 0.8 μA at 2000 K.

Practically, to obtain a permanent emission current, one must apply an external electric field : in the absence of such a field, the first electrons emitted into vacuum create a space charge, associated with an antagonistic electric field, which prevents any further emission.

Another way of calculating the emitted current consists in considering that in the TV tube, in the absence of applied electrical potential, an equilibrium sets up at temperature T between the tungsten electrons which try to escape the metal and those outside the metal that try to enter it.

The equilibrium between electrons in both phases is expressed through the equality of their chemical potentials. Now, outside the solid the electron concentration is very small, so that one deals with a classical gas which follows the Maxwell-Boltzmann statistics. The electron current toward the solid is calculated like for an ideal gas. When an electric field is applied, this current vanishes but the emission current, which is its opposite, remains.

Note that, in order to be rigorous, all these calculations should take into account the energy levels of the metal and its surface, which would be very complicated! Anyhow, the above estimates qualitatively describe the effect and lead to correct orders of magnitude.

On this example of the thermionic emission we have shown a practical application, in which electrons follow the high temperature limit of the Fermi-Dirac statistics.

Summary of Chapter 7

The properties of free fermions are governed by the Fermi-Dirac distribution :

$$f_{FD}(\varepsilon) = \frac{1}{\exp \beta(\varepsilon - \mu) + 1}$$

At zero temperature the fermions occupy all the levels of energy lower or equal to ε_F, the Fermi energy, which is defined as the chemical potential μ for $T = 0$ K.

The number N of particles of the system and the Fermi energy are linked at $T = 0$ K through

$$N = \int_0^{\varepsilon_F} D(\varepsilon)d\varepsilon$$

This situation, related to the Pauli principle, induces large values of the internal energy and of the pressure of a system of N fermions.

An excitation, of thermal or magnetic origin, of a fermions system, only affects the states of energy close to μ or ε_F. Consequently, the fermions' magnetic susceptibility and specific heat are reduced with respect to their values in a classical system. In the case of the specific heat, the reduction is a ratio of the order of $k_B T/\mu$, where μ is the chemical potential, of energy close to ε_F.

The magnetic susceptibility and the specific heat of electrons in metals are well described in the framework of the free electrons model developed in this chapter.

Appendix 7.1

Calculation of C_v Through the Sommerfeld Development

The specific heat $C_v = \left(\dfrac{dU}{dT}\right)_{N,\Omega}$ of fermions is deduced from the expression of the internal energy $U(T)$ versus temperature.

We have therefore to calculate, at nonvanishing temperature,

$$N = \int_0^\infty f(\varepsilon)D(\varepsilon)d\varepsilon \tag{7.36}$$

$$U = \int_0^\infty \varepsilon f(\varepsilon)D(\varepsilon)d\varepsilon \tag{7.37}$$

that is, more generally,

$$\langle g(\varepsilon)\rangle = \int_{-\infty}^{+\infty} g(\varepsilon)D(\varepsilon)f(\varepsilon)d\varepsilon = \int_{-\infty}^{+\infty} \varphi(\varepsilon)f(\varepsilon)d\varepsilon \tag{7.38}$$

where $\varphi(\varepsilon)$ is assumed to vary like a power law of ε for $\varepsilon > \varepsilon_{\text{minimum}}$, to be null for smaller values of ε.

We know that, owing to the shape of the Fermi distribution,

$$\lim_{T\to 0}\langle g(\varepsilon)\rangle = \int_{-\infty}^{\varepsilon_F} \varphi(\varepsilon)d\varepsilon \tag{7.39}$$

We write the difference

$$\Delta = \int_{-\infty}^{+\infty} \varphi(\varepsilon)f(\varepsilon)d\varepsilon - \int_{-\infty}^{\mu} \varphi(\varepsilon)d\varepsilon \tag{7.40}$$

which is small when $k_B T \ll \mu$ ($\beta\mu \gg 1$). This difference introduces $\varphi'(\mu)$: if $\varphi'(\mu)$ is zero, because of the symmetry of $f(\varepsilon)$ around μ, the contributions on

171

both sides of μ compensate. Besides, we understood in § 7.2.1 that, to maintain the same N when the temperature increases, μ gets closer to the origin, for a three-dimensional system for which $\varphi'(\mu) > 0$ when $D(\varepsilon)$ is different from zero ($\varphi(\varepsilon) \propto \sqrt{\varepsilon}$).

Let us write Δ explicitly :

$$\Delta = \int_{-\infty}^{\mu} d\varepsilon \varphi(\varepsilon) f(\varepsilon) - \int_{-\infty}^{\mu} d\varepsilon \varphi(\varepsilon) + \int_{\mu}^{\infty} d\varepsilon \varphi(\varepsilon) f(\varepsilon) \qquad (7.41)$$

$$= \int_{-\infty}^{\mu} d\varepsilon \varphi(\varepsilon) \left(\frac{1}{e^{\beta(\varepsilon-\mu)} + 1} - 1 \right) + \int_{\mu}^{\infty} d\varepsilon \varphi(\varepsilon) \frac{1}{e^{\beta(\varepsilon-\mu)} + 1} \qquad (7.42)$$

Here a symmetrical part is played by either the empty states below μ [the probability for a state to be empty is $1 - f(\varepsilon)$], or the filled states above μ, all in small numbers if the temperature is low.

We take $\beta(\varepsilon - \mu) = x$, $\beta d\varepsilon = dx$

$$\Delta = - \int_{-\infty}^{0} \frac{dx}{\beta} \varphi \left(\mu + \frac{x}{\beta} \right) \frac{1}{1 + e^{-x}} + \int_{0}^{\infty} \frac{dx}{\beta} \varphi \left(\mu + \frac{x}{\beta} \right) \frac{1}{1 + e^{x}} \qquad (7.43)$$

$$= - \int_{0}^{\infty} \frac{dx'}{\beta} \varphi \left(\mu - \frac{x'}{\beta} \right) \frac{1}{1 + e^{x'}} + \int_{0}^{\infty} \frac{dx}{\beta} \varphi \left(\mu + \frac{x}{\beta} \right) \frac{1}{1 + e^{x}} \qquad (7.44)$$

$$= \int_{0}^{\infty} \frac{dx}{\beta} \frac{1}{1 + e^{x}} \left[\varphi \left(\mu + \frac{x}{\beta} \right) - \varphi \left(\mu - \frac{x}{\beta} \right) \right] \qquad (7.45)$$

$$= \frac{1}{\beta} \int_{0}^{\infty} \frac{dx}{1 + e^{x}} \cdot 2 \left[\varphi'(\mu) \frac{x}{\beta} + \frac{\varphi^{(3)}(\mu)}{3!} \left(\frac{x}{\beta} \right)^3 + \dots \right] \qquad (7.46)$$

$$= \sum_{n=0}^{\infty} \frac{\varphi^{(2n+1)}(\mu)}{\beta^{2n+2}} I_n \qquad (7.47)$$

with

$$I_n = \frac{2}{(2n+1)!} \int_{0}^{\infty} dx \frac{x^{2n+1}}{e^x + 1} \qquad (7.48)$$

The integrals I_n are deduced from the Riemann ζ functions (see the section "A few useful formulae"). In particular at low temperature the principal term of Δ is equal to $\frac{\pi^2}{6}(k_B T)^2 \varphi'(\mu)$.

Let us return to the calculation of the specific heat C_v. For the internal energy of nonrelativistic particles, $\varphi(\varepsilon) = \varepsilon \cdot K\varepsilon^{1/2} = K\varepsilon^{3/2}$ for $\varepsilon > 0$, $\varphi(\varepsilon) = 0$ for $\varepsilon < 0$. Then

$$U(T) = \int_{0}^{\mu} K\varepsilon^{3/2} d\varepsilon + \frac{\pi^2}{6}(k_B T)^2 (K\varepsilon^{3/2})'_{\mu} + O(T^4) \qquad (7.49)$$

but μ depends on T since

$$N = \int_0^{\mu} K\varepsilon^{1/2} d\varepsilon + \frac{\pi^2}{6}(k_B T)^2 (K\varepsilon^{1/2})'_{\mu} + O(T^4) = \int_0^{\varepsilon_F} K\varepsilon^{1/2} d\varepsilon \quad (7.50)$$

expresses the conservation of the particles number.

One deduces from (7.50) :

$$\mu(T) = \varepsilon_F \left[1 - \frac{\pi^2}{12}\left(\frac{k_B T}{\varepsilon_F}\right)^2 \right] + O(T^4) \quad (7.51)$$

Replacing the expression of $\mu(T)$ into $U(T)$, one gets :

$$U(T) - U(T=0) \simeq K\varepsilon_F^{5/2}\frac{\pi^2}{6}\left(\frac{k_B T}{\varepsilon_F}\right)^2 = N\varepsilon_F \frac{\pi^2}{4}\left(\frac{k_B T}{\varepsilon_F}\right)^2 \quad (7.52)$$

One deduces C_v :

$$C_v = \frac{dU}{dT} = \frac{3}{2}Nk_B \cdot \frac{\pi^2}{3}\left(\frac{k_B T}{\varepsilon_F}\right) \quad (7.53)$$

Chapter 8

Elements of Bands Theory and Crystal Conductivity

The previous chapter on free fermions allowed us to interpret some properties of metals. We considered that each electron was confined within the solid by potential barriers, but did not introduce any specific description of its interactions with the lattice ions or with the other electrons. This model is not very realistic, in particular it cannot explain why some solids are conducting the electrical current and others are not : at room temperature, the conductivity of copper is almost 10^8 siemens ($\Omega^{-1} \cdot m^{-1}$) whereas that of quartz is of 10^{-15} S, and of teflon 10^{-16} S. This physical parameter varies on 24 orders of magnitude according to the material, which is absolutely exceptional in nature.

Thus we are going to reconsider the study of electrons in solids from the start and to first address the characteristics of a solid and of its physical description. As we now know, the first problem to solve is a quantum mechanical one, and we will rely on your previous courses. In particular we will be concerned by a periodic solid, that is, a crystal (§ 8.1). We will see that the electron eigenstates are regrouped into energy bands, separated by energy gaps (§ 8.2 and § 8.3).

We will have to introduce here very simplifying hypotheses : in Solid State Physics courses, more general assumptions are used to describe electronic states of crystals.

Using Statistical Physics (§ 8.4) we will see how electrons are filling energy bands. We will understand why some materials are conductors, other ones insulators. We will go more into details on the statistics of semiconductors, of high importance in our everyday life, and will end by a brief description of the principles of a few devices based on semiconductors (§ 8.5).

8.1 What is a Solid, a Crystal ?

A solid is a dense system with a specific shape of its own. This definition expresses mechanical properties; but one can also refer to the solid's visual aspect, shiny in the case of metals, transparent in the case of glass, colored for painting pigments. These optical properties, as well as the electrical conduction and the magnetic properties, are due to the electrons. In a macroscopic solid with a volume in the cm^3 range, the numbers of electrons and nuclei present are of the order of the Avogadro number $\mathcal{N} = 6.02 \times 10^{23}$.

The hamiltonian describing this solid contains for each particle, either electron or nucleus, its kinetic energy, and its potential energy arising from its Coulombic interaction with all the other charged particles. It is easily guessed that the exact solution will be impossible to get without approximations (and still very difficult to obtain using them!).

One first notices that nuclei are much heavier than electrons and, consequently, much less mobile : either the nuclei motion will be neglected by considering that their positions remain fixed, or the small oscillations of these nuclei around their equilibrium positions (solid vibrations) will be treated separately from the electron motions, so that the electron hamiltonian $\hat{H}$ will not include the nuclei kinetic energy. $\hat{H}$ will then be given by

$$\hat{H} = \sum_{i=1}^{NZ} \frac{\hat{p}_i^2}{2m} - \frac{Ze^2}{4\pi\varepsilon_0} \sum_{i=1}^{NZ} \sum_{n=1}^{N} \frac{1}{|\vec{r}_i - \vec{R}_n|} + \frac{1}{2}\frac{e^2}{4\pi\varepsilon_0} \sum_{i,j=1}^{NZ} \frac{1}{|\vec{r}_i - \vec{r}_j|} \qquad (8.1)$$

where Z is the nuclear charge; the nth nucleus is located at $\vec{R}_n$ and the electron i at $\vec{r}_i$. The second term expresses the Coulombian interaction of all the electrons with all the nuclei, the third one the repulsion among electrons.

Besides, we will assume that the solid is a perfect crystal, that is, its atoms are regularly located on the sites $\vec{R}_n$ of a three-dimensional periodic lattice. This will simplify the solution of the Quantum Mechanical problem, but many of our conclusions will also apply to disordered solids.

The repulsion term between electrons in $\hat{H}$ depends of their probabilities of presence, in turn given by the solution of the hamiltonian : this is a difficult self-consistent problem. The Hartree method, which averages the electronic repulsion, and the Hartree-Fock method, which takes into account the wave function antisymmetrization as required by the Pauli principle, are used to solve this problem by iteration, but they lead to complicated calculations. Phenomena like ferromagnetism, superconductivity, Mott insulators, mixed-valence systems, are based on the interaction between electrons; however in most cases one can account of the repulsion between electrons in a crystal through an *average* one-electron periodic term and we will limit ourselves to this situation in the present course. Then it will be possible to split the hamiltonian $\hat{H}$ into one-electron terms, this is the *independent-electron* approximation :

$$\hat{H} = \sum_{i=1}^{N} \hat{h}_i \qquad (8.2)$$

with

$$\hat{h}_i = \frac{\hat{p}_i^2}{2m} + \mathcal{V}(\vec{r}_i) \qquad (8.3)$$

The potential $\mathcal{V}(\vec{r}_i)$ has the lattice periodicity, it takes into account the attractions and repulsions on electron i and is globally attractive. The eigenvalues of $\hat{H}$ are the sums of the eigenenergies of the various electrons.

Our task will consist in solving the one-electron hamiltonian $\hat{h}_i$. We will introduce an additional simplification by assuming that the lattice is one-dimensional, i.e., $\mathcal{V}(\vec{r}_i) = \mathcal{V}(x_i)$. This will allow us to extract the essential characteristics, at the cost of simpler calculations.

In the "electron in the box" model that we have used up to now, $\mathcal{V}(x_i)$ only expressed the potential barriers limiting the solid, and thus leveled out the potential of each atom. Here we will introduce the attractive potential on each lattice site. The energy levels will derive from atomic levels, rather than from the free electron states which were the solutions in the "box" case.

8.2 The Eigenstates for the Chosen Model

8.2.1 Recall : the Double Potential Well

First consider an electron on an *isolated atom*, located at the origin. Its hamiltonian $\hat{h}$ is given by

$$\hat{h} = \frac{\hat{p}^2}{2m} + V(x) \qquad (8.4)$$

where V is the even Coulombian potential of the nucleus (Fig. 8.1).

One of its eigenstate is such that :

$$\hat{h}\varphi(x) = \varepsilon_0\varphi(x) \tag{8.5}$$

$\varphi(x)$ is an atomic orbital, of s-type for example, which decreases from the origin with a characteristic distance λ. We will assume that the other orbitals do not have to be considered in this problem, since they are very distant in energy from ε_0.

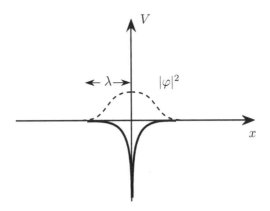

FIG. 8.1: Coulombian potential of an isolated ion and probability of presence of an electron on this ion.

Let us add another nucleus, that we locate at $x = d$ (Fig. 8.2). The hamiltonian, still for a *single* electron now sharing its presence *between the two ions*,

$$\hat{h} = \frac{\hat{p}^2}{2m} + V(x) + V(x - d) \tag{8.6}$$

In the case of two $1s$-orbital initial states, this corresponds to the description of the H_2^+ ion (two protons, one electron).

One follows the standard method on the double potential well of the Quantum Mechanics courses and defines $|\varphi_L\rangle$ and $|\varphi_R\rangle$, the atomic eigenstates respectively located on the left nucleus (at $x = 0$) and the right nucleus (at $x = d$) :

$$\begin{cases} \left[\dfrac{\hat{p}^2}{2m} + V(x)\right]|\varphi_L\rangle & = \varepsilon_0|\varphi_L\rangle \\[4mm] \left[\dfrac{\hat{p}^2}{2m} + V(x - d)\right]|\varphi_R\rangle & = \varepsilon_0|\varphi_R\rangle \end{cases} \tag{8.7}$$

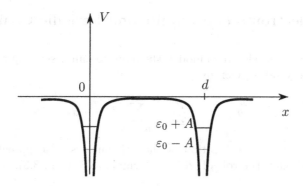

FIG. 8.2: Potential due to two ions of the same nature, distant of d and corresponding bound states.

Owing to the symmetry of the potential in (8.6) with respect to the point $x = d/2$, whatever d the eigenfunctions of $\hat{h}$ can be chosen either symmetrical or antisymmetrical with respect to this mid-point. However, if d is very large with respect to λ, the coupling between the two sites is very weak, $\hat{h}$ has two eigenvalues practically equal to ε_0 and if at the initial time the electron is on one of the nuclei, for example the left one L, it will need a time overcoming the observation possibilities to move to the right one R. One can then consider that the atomic states (8.7) are the eigenstates of the problem (although strictly they are not).

On the other hand, you learnt in Quantum Mechanics that if d is not too large with respect to λ, the L and R wells are coupled by the tunnel effect, which is expressed through the coupling matrix element

$$-A = \langle \varphi_R | \hat{H} | \varphi_L \rangle \quad , \quad A > 0 \tag{8.8}$$

which decreases versus d approximately like $\exp(-d/\lambda)$.

The energy degeneracy is then lifted : the two new eigenvalues of $\hat{h}$ are equal to $\varepsilon_0 - A$, corresponding to the symmetrical wave function $|\psi_S\rangle = \frac{1}{\sqrt{2}}(|\varphi_L\rangle + |\varphi_R\rangle)$, and $\varepsilon_0 + A$, of antisymmetrical wave function $|\psi_A\rangle = \frac{1}{\sqrt{2}}(|\varphi_L\rangle - |\varphi_R\rangle)$ (Fig. 8.2). These results are valid for a weak enough coupling ($A \ll |\varepsilon_0|$).

For a stronger coupling ($d \sim \lambda$) one should account for the overlap of the two atomic functions, expressed by $\langle \varphi_L | \varphi_R \rangle$ different from zero. Anyway the eigenstates are delocalized between both wells. For H_2^+ they correspond to the bonding and antibonding states of the chemical bound.

8.2.2   Electron on an Infinite and Periodic Chain

In the independent electrons model, the hamiltonian describing a *single* electron in the crystal is given by

$$\hat{h}_{\text{crystal}} = \frac{\hat{p}^2}{2m} + \sum_{n=-\infty}^{+\infty} V(x - x_n) \tag{8.9}$$

where $x_n = nd$, is the location of the nth nucleus. The potentials sum is periodic and plays the role of $\mathcal{V}(\vec{r}_i)$ in formula (8.3) (Fig. 8.3).

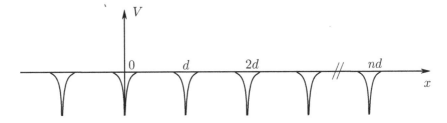

FIG. 8.3 : Potential periodic of a crystal.

One is looking for the stationary solution, such that

$$\hat{h}_{\text{crystal}}\psi(x) = \varepsilon\psi(x) \tag{8.10}$$

If the nuclei are far enough apart, there is no coupling : although the eigenstates are delocalized, if at the initial time the electron is on a given nucleus, an infinite time would be required for this electron to move to the neighboring nucleus. The state of energy $\varepsilon = \varepsilon_0$ is N times degenerate. On the other hand, in the presence of an even weak coupling, the eigenstates are delocalized. We are going to assume that the coupling is weak and will note :

$$\hat{h}_{\text{crystal}} = \hat{h}_n + \sum_{n' \neq n} V(x - n'd) \tag{8.11}$$

to show the atomic hamiltonian on the specific site n.

To solve the eigenvalue problem, by analogy with the case of the double potential well, we will work in a basis of localized states, which are not the eigenstates of the considered infinite periodic chain. Now every site plays a similar role, the problem is invariant under the translations which change x into $x + nd = x + x_n$. The general basis state is written $|n\rangle$, it corresponds to the wave function $\phi_n(x)$ located around x_n and is deduced through the translation x_n from the function centered at the origin, i.e., $\phi_n(x) = \phi_0(x - x_n)$.

These states are built through orthogonalization from the atomic functions $\varphi(x - x_n)$ in order to constitute an orthonormal basis : $\langle n'|n\rangle = \delta_{nn'}$ (this is the so-called "Hückel model"). The functions $\phi_n(x)$ depend on the coupling. When the atoms are very distant, $\phi_n(x)$ is very close to $\varphi(x - x_n)$, the atomic wave function. In a weak coupling situation, where the tunnel effect only takes place between first neighbors, $\phi_n(x)$ is still not much different from the atomic function.

One looks for a stationary solution of the form

$$|\psi\rangle = \sum_n c_n |n\rangle \tag{8.12}$$

One needs to find the coefficients c_n, expected to all have the same norm, since every site plays the same role :

$$\hat{h}_{\text{crist}}\left(\sum_{n=-\infty}^{+\infty} c_n |n\rangle \right) = \varepsilon \left(\sum_{n=-\infty}^{+\infty} c_n |n\rangle \right) \tag{8.13}$$

One proceeds by analogy with the solution of the double-well potential problem : one assumes that the hamiltonian only couples the first neighboring sites. Then the coupling term is given by

$$\langle n | \hat{h}_{\text{crist}} | n + 1 \rangle = \langle n - 1 | \hat{h}_{\text{crist}} | n \rangle = -A \tag{8.14}$$

For atomic s-type functions $\varphi(x)$ A is negative.

By multiplying (8.13) by the bra $\langle n |$, one obtains

$$c_{n-1}\langle n | \hat{h}_{\text{cryst}} | n - 1 \rangle + c_n \langle n | \hat{h}_{\text{cryst}} | n \rangle + c_{n+1} \langle n | \hat{h}_{\text{cryst}} | n + 1 \rangle = c_n \varepsilon \tag{8.15}$$

Now

$$\langle n | \hat{h}_{\text{cryst}} | n \rangle = \langle n | \hat{h}_n | n \rangle + \langle n | \sum_{n' \neq n} V(x - n'd) | n \rangle$$
$$= \varepsilon_0 + \alpha \simeq \varepsilon_0 \tag{8.16}$$

The term α, which is the integral of the product of $|\phi_n(x)|^2$ by the sum of potentials of sites distinct from x_n, is very small in weak coupling conditions [it varies in $\exp(-2d/\lambda)$, whereas A is in $\exp(-d/\lambda)$]. It will be neglected with respect to ε_0.

The coefficients c_n, c_{n-1} and c_{n+1} are then related through

$$-c_{n-1}A + c_n \varepsilon_0 - c_{n+1}A = c_n \varepsilon \tag{8.17}$$

One obtains analogous equations by multiplying (8.13) by the other bras, whence finally the set of coupled equations :

$$\begin{cases} \cdots\cdots \\ -c_{n-1}A + c_n \varepsilon_0 - c_{n+1}A \qquad\qquad = c_n \varepsilon \\ \qquad\quad - c_n A + c_{n+1}\varepsilon_0 - c_{n+2}A \;\; = c_{n+1}\varepsilon \end{cases} \tag{8.18}$$

8.2.3   Energy Bands and Bloch Functions

The system (8.18) assumes a nonzero solution if one chooses c_n as a phase :

$$c_n = \exp(ik \cdot nd) \tag{8.19}$$

where k, homogeneous to a reciprocal length, is a wave vector and nd the coordinate of the considered site.

By substituting (8.19) into (8.17) one obtains the dispersion relation, or dispersion law, linking ε and k :

$$\varepsilon(k) = \varepsilon_0 - 2A \cos kd \tag{8.20}$$

$\varepsilon(k)$ is an even function, periodic in k, which describes the totality of its values, i.e. the interval $[\varepsilon_0 - 2A, \varepsilon_0 + 2A]$, when k varies from $-\pi/d$ to π/d (Fig. 8.4). The domain of accessible energies ε constitutes an *allowed band* ; a value of ε chosen outside of the interval $[\varepsilon_0 - 2A, \varepsilon_0 + 2A]$ does not correspond to a k-value, it belongs to a *forbidden* domain. The range $-\pi/d \leq k < \pi/d$ is called "the first Brillouin zone," a concept introduced in 1930 by Léon Brillouin, a French physicist ; it is sufficient to study the physics of the problem. Note that

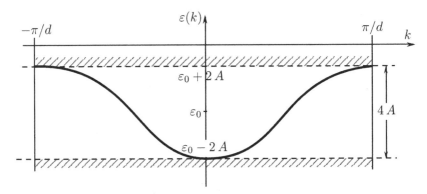

FIG. 8.4 : Dispersion relation in the crystal.

in the neighborhood of $k = 0$, $\varepsilon(k) \simeq \varepsilon_0 - 2A + Ad^2 k^2 + \ldots$ which looks like the free electron dispersion relation, if an *effective mass* m^* is defined such that $\hbar^2/2m^* = Ad^2$. This mass also expresses the smaller or larger difficulty for an electron to move under the application of an external electric field, due to the interactions inside the solid it is subjected to.

The wave function associated with $\varepsilon(k)$ is given by

$$\psi_k(x) = \sum_{n=-\infty}^{+\infty} \exp iknd \cdot \phi_0(x - nd) \tag{8.21}$$

This expression can be transformed into

$$\psi_k(x) = \exp ikx \cdot \sum_{n=-\infty}^{+\infty} \exp -ik(x - nd) \cdot \phi_0(x - nd) \qquad (8.22)$$

which is the product of a plane wave by a periodic function $u_k(x)$, of period d : indeed, for n' integer,

$$u_k(x + n'd) = \sum_{n=-\infty}^{+\infty} \exp -ik(x + n'd - nd) \cdot \phi_0(x + n'd - nd)$$

$$= \sum_{n''=-\infty}^{+\infty} \exp -ik(x - n''d) \cdot \phi_0(x - n''d) = u_k(x) \qquad (8.23)$$

This type of solution of the Schroedinger stationary equation

$$\psi_k(x) = e^{ikx} \cdot u_k(x) \qquad (8.24)$$

can be generalized, for a three-dimensional periodic potential $V(\vec{r})$, into

$$\psi_{\vec{k}}(\vec{r}) = e^{i\vec{k}\cdot\vec{r}} \cdot u_{\vec{k}} \qquad (8.25)$$

This is the so-called *Bloch function*, from the name of Felix Bloch, who, in his doctoral thesis prepared in Leipzig under the supervision of Werner Heisenberg and defended in July 1928, was the first to obtain the expression of a crystal wave function, using a method analogous to the one followed here. The expression results from the particular form of the differential equation to be solved (second derivative and periodic potential) and is justified using the Floquet theorem.

Expression (8.21) of $\psi_k(x)$, established in the case of weak coupling between neighboring atomic sites, can be interpreted as a *Linear Combination of Atomic Orbitals* (LCAO), since the ϕ_0's then do not differ much from the atomic solutions φ. This model is also called *"tight binding"* as it privileges the atomic states. When the atomic states of two neighboring sites are taken as orthogonal, this is the so-called "Hückel model."

The $\psi_k(x)$ functions are delocalized all over the crystal. The extreme energy states, at the edges of the energy band, correspond to particular Bloch functions : the minimum energy state $\varepsilon_0 - 2A$ is associated with $\psi_0(x) = \sum_{n=-\infty}^{+\infty} \phi_0(x-nd)$, the addition in phase of each localized wavefunction : this is a situation comparable to the bonding symmetrical state of the H_2^+ ion ; the state of energy $\varepsilon_0 + 2A$ is associated with $\psi_{\pi/d}(x) = \sum_{n=-\infty}^{+\infty}(-1)^n\phi_0(x-nd)$, alternate addition of each localized wave function and is analogous to the antibonding state of the H_2^+ ion.

One can notice that the band width $4A$ obtained for an infinite chain is only twice the value of the energy splitting for a pair of coupled sites. The reason is that, in the present model, a given site only interacts with its first neighbors and not with the more distant sites

Note : the dispersion relation (8.20) and the wave function (8.24), that we obtained under very specific hypotheses, can be generalized in the case of three-dimensional periodic solids, as can be found in basic Solid State Physics courses, like the one taught at Ecole Polytechnique.

8.3   The Electron States in a Crystal

8.3.1   Wave Packet of Bloch Waves

Bloch waves of the (8.21) or (8.24) type are delocalized and thus cannot describe the amplitude of probability of presence for an electron, as these waves cannot be normalized : although a localized function $\phi_0(x)$ can be normed, $\psi_k(x)$ spreads to infinity.

To describe an electron behavior, one has to build a packet of Bloch waves

$$\Psi(x,t) = \int_{-\infty}^{+\infty} g(k)e^{ikx}u_k(x)e^{-i\varepsilon(k)t/\hbar}dk \qquad (8.26)$$

centered around a wave vector k_0, of extent Δk small with respect to π/d. From the Heisenberg uncertainty principle, the extent in x of the packet will be such that $\Delta x \cdot \Delta k \geq 1$, i.e., $\Delta x \gg d$: the Bloch wave packet is spreading over a large number of elementary cells.

8.3.2   Resistance ; Mean Free Path

It can be shown that an electron described by a wave packet (8.26), in motion in a *perfect infinite* crystal, keeps its velocity value for ever and thus is not diffused by the periodic lattice ; in fact the Bloch wave solution already includes all these coherent diffusions. Then there is absolutely *no resistance* to the electron motion or, in other words, the conductivity is infinite.

A static applied electric field $\vec{E}$ modifies the electron energy levels ; the problem is complex, one should also account for the effect of the lattice ions which are also subjected to the action of $\vec{E}$ and interact with the electron. Experimentally one notices in a macroscopic crystal that the less its defects, like dislocations or impurities, the higher its conductivity. An infinite perfect three-dimensional crystal would have an infinite conductivity, a result predicted by Felix Bloch as early as 1928. (These simple considerations do not apply to the recently elaborated, low-dimensional, systems.)

The real finite conductivity of crystals arises from the discrepancies to the perfect infinite lattice : thermal motion of the ions, defects, existence of the surface. In fact in copper at low temperature one deduces from the conductivity value $\sigma = ne^2\tau/m$ the mean time τ between two collisions and, introducing the Fermi velocity v_F (see §7.1.1), one obtains a mean free path $\ell = v_F\tau$ of the order of 40 nm, that is, about 100 interatomic distances. This does show that the lattice ions are not responsible for the diffusion of conduction electrons.

8.3.3 Finite Chain, Density of States, Effective Mass

FIG. 8.5 : Potential for a N-ions chain.

A real crystal is spreading over a macroscopic distance $L = Nd$ (Fig. 8.5) (the distance L is measured with an uncertainty much larger than the interatomic distance d; one can thus always assume that L contains an integer number N of interatomic distances). Obviously, if only the coupling to the nearest neighbors is considered, for most sites there is no change with respect to the case of the infinite crystal. Thus the eigenstates of the hamiltonian for the finite crystal :

$$\hat{h}_{\text{crystal}} = \frac{\hat{p}^2}{2m} + \sum_{n=1}^{N} V(x - x_n) \tag{8.27}$$

should not differ much from those of the hamiltonian (8.9) for the infinite crystal : we are going to test the Bloch function obtained for the infinite crystal as a solution for the finite case.

On the other hand, for a macroscopic dimension L the exact limit conditions do not matter much. It is possible, as in the case of the free electron model, to close the crystal on itself by superposing the sites 0 and Nd and to express the periodic limit conditions of Born-Von Kármán :

$$\psi(x + L) = \psi(x) \tag{8.28}$$

$$\psi_k(x + L) = e^{ik(x+L)}u_k(x + L) = e^{ikL}\psi_k(x) \tag{8.29}$$

as L contains an integer number of atoms, i.e. an integer number of atomic periods d, and $u_k(x)$ is periodic. One thus finds the *same quantization* condition as for a one-dimensional *free particle* : $k = p2\pi/L$, where p is an integer, associated with an one-dimension orbital density of states $L/2\pi$; the corresponding three-dimensional density of states is $\Omega/(2\pi)^3$, where Ω is the crystal volume. These densities are doubled when one accounts for the electron spin.

The solutions of (8.27) are then

$$\begin{cases} \varepsilon = \varepsilon_0 - 2A\cos kd \\ \psi_k(x) = \frac{1}{\sqrt{N}}\sum_{n=1}^{N}\exp iknd \cdot \phi_0(x - nd) \\ \text{with } k = p \cdot \frac{2\pi}{L} \qquad p \text{ integer} \end{cases} \tag{8.30}$$

When comparing to the results (8.20) and (8.21) for the infinite crystal, one first notices that $\psi_k(x)$ as given by (8.30) is normed. Moreover ε is now quantized (Fig. 8.6) : to describe the first Brillouin zone $\left(-\dfrac{\pi}{d} \leq k < \dfrac{\pi}{d}\right)$ all the different states are obtained for p ranging between $-\dfrac{N}{2}$ and $+\dfrac{N}{2} - 1$ (or between $-\dfrac{N-1}{2}$ and $\dfrac{N-1}{2} - 1$ if N is odd, but because of the order of magnitude of N, ratio of the macroscopic distance L to the atomic one d, this does not matter much !). On this interval there are thus N possible states, i.e., as many as the number of sites. Half of the states correspond to waves propagating toward increasing x's ($k > 0$), the other ones to waves propagating toward decreasing x's.

Note that if the states are uniformly spread in $\vec{k}$, their distribution is not uniform in energy. One defines the energy density of states $D(\varepsilon)$ such that the number of accessible states dn, of energies between ε and $\varepsilon + d\varepsilon$, is equal to

$$dn = D(\varepsilon)d\varepsilon \tag{8.31}$$

One will note that, at one dimension, to a given value of ε correspond two opposite values of $\vec{k}_x$, associated with distinct quantum states. By observing Fig. 8.6, it can be seen that $D(\varepsilon)$ takes very large values at the vicinity of the band edges (ε close to $\varepsilon_0 \pm 2A$) ; in fact at *one* dimension $D(\varepsilon)$ has a specific behavior [$D(\varepsilon)$ diverges at the band edges, but the integral

$$\int_{\varepsilon_{min}}^{\varepsilon} D(\varepsilon')d\varepsilon' \tag{8.32}$$

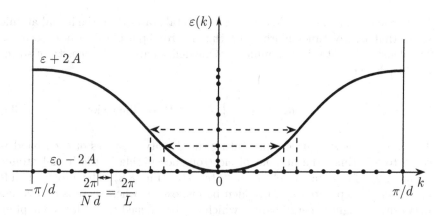

FIG. 8.6 : Allowed states in the energy band, for a N electrons system.

remains finite]. When this model is extended to a three-dimensional system, one finds again a "quasicontinuous" set of very close accessible states, uniformly located in $\vec{k}$ (but not in energy), in number equal to that of the initial orbitals (it was already the case for two atoms for which we obtained the bonding and antibonding states).

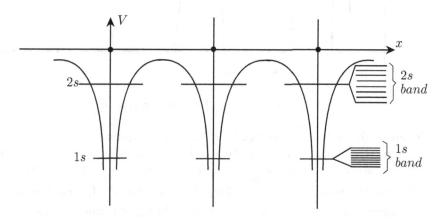

FIG. 8.7: In a N atom crystal, to each atomic state corresponds a band that can accommodate $2N$ times more electrons than the initial state.

In a more realistic case, the considered atom has several energy levels $(1s, 2s, 2p, ...)$. Each energy level gives rise to a band, wider if the atomic state lies higher in energy (Fig. 8.7). The widths of the bands originating from the deep atomic states are extremely small, one can consider that these atomic levels do not couple. On the other hand, the width of the band containing the electrons involved in the chemical bond is always of the order of several electron-volts.

Again each band contains N times more orbital states than the initial atomic state, that is, $2N$ times more when the electron spin is taken into account. The density of states in $\vec{k}$ is uniform. One defines an energy density of states $D(\varepsilon)$ such that

$$dn = D(\varepsilon)d\varepsilon = 2\left(\frac{L}{2\pi}\right)^3 d^3\vec{k} \text{ at three dimensions} \tag{8.33}$$

The area under the curve representing the density of states of each band is equal to $2N$ times the number of electrons acceptable in the initial atomic orbital : $2N$ for s states, $6N$ for p states, etc. (Fig. 8.8). The allowed bands are generally separated by forbidden bands, except when there is an overlap between the higher energy bands, which are broadened through the coupling ($3s$ and $3p$ on the Fig. 8.8).

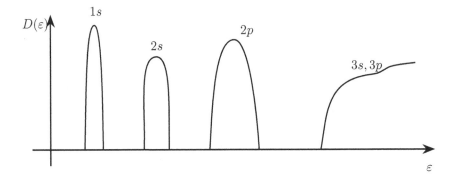

FIG. 8.8 : Density of states of a crystal.

It can be shown that in the vicinity of a band extremum, the tangent of the $D(\varepsilon)$ curve is vertical. In particular, at the bottom of a band the density of state is rather similar to the one of a free electron in a box [see (6.64)], for which we obtained $D(\varepsilon) = \dfrac{\Omega}{2\pi^2\hbar^3}(2m)^{3/2}\sqrt{\varepsilon}$, m being the free electron mass. Here again one will define the *effective mass* in a given band, this time from the curvature of $D(\varepsilon)$ [this mass is the same as the one in § 8.2.3, introduced from the dispersion relation (8.20)]. Indeed, setting

$$D(\varepsilon) \simeq \frac{\Omega}{2\pi^2\hbar^3}(2m^*)^{3/2}\sqrt{\varepsilon - \varepsilon_0} \tag{8.34}$$

in the vicinity of ε_0 at the bottom of the band is equivalent to taking for dispersion relation

$$\varepsilon - \varepsilon_0 \simeq \frac{\hbar^2 k^2}{2m^*} \quad , \quad \text{i.e.,} \quad \frac{\hbar^2}{m^*} = \frac{d^2\varepsilon}{dk^2} \tag{8.35}$$

The effective mass is linked to the band width : from the dispersion relation (8.20) for instance, one obtains around $k = 0$

$$\frac{1}{m^*} = \frac{2Ad^2}{\hbar^2} \tag{8.36}$$

One can understand that the electrons of the deep bands are "very heavy," and thus not really mobile inside the lattice, that is, they are practically localized.

Note : in non crystalline (or amorphous) solids, like amorphous silicon, the coupling between the atomic states also produces energy bands.

8.4 Statistical Physics of Solids

Once the one-electron quantum states determined, one has to locate the N electrons in these states, in the framework of the independent electrons model.

8.4.1 Filling of the Levels

At zero temperature, the Pauli principle dictates one to fill the levels from the lowest energy one, putting two electrons, with different spin states, on each orbital state until the N electrons are exhausted. We now take two examples, that of lithium ($Z = 3$), and that of carbon ($Z = 6$) crystallized into the diamond form.

In lithium, each atom brings 3 electrons, for the whole solid there are $3N$ electrons to be located. The $1s$ band is filled by $2N$ electrons. The complementary N electrons are filling the $2s$ band up to its half, until the energy $\varepsilon = \varepsilon_F$, the Fermi energy, which lies 4.7 eV above the bottom of this band. Above ε_F, any state is empty at $T = 0$ K.

When a small amount of energy is brought to the solid, for example using a voltage source of a few volts, the electrons in the $2s$ band in the vicinity of ε_F can occupy states immediately above ε_F, whereas $1s$ electrons cannot react, having no empty states in their neighborhood. This energy modification of electrons in the $2s$ band produces a global macroscopic reaction, expressed by an electric current, to the electric field : lithium is a *metal* or a conductor, the $2s$ band is its *conduction band*.

In diamond, the initial electronic states are not the $2s$ and $2p$ states of the atomic carbon but their sp^3 tetrahedral hybrids (as in CH_4) which separate into bonding hybrids with 4 electrons per atom (including spin) and antibonding hybrids also with 4 electrons. In the crystal, the bonding states constitute a $4N$-electron band, the same is true for the antibonding states. The $4N$ electrons of the N atoms exactly fill the bonding band, which contains the electrons participating in the chemical bond, whence the name of *valence band*. The band immediately above, of antibonding type, is the *conduction band*. It is separated from the valence band by the *band gap* energy, noted E_g : $E_g = 5.4$ eV for diamond and at $T = 0$ K the conduction band is empty, the valence band having exactly accommodated all the available electrons. Under the effect of an applied voltage of a few volts, the valence electrons cannot be excited through the forbidden band, there is no macroscopic reaction, the diamond is an *insulator*.

Like carbon, silicon and germanium also belong to column IV of the periodic table, they have the same crystalline structure, they are also insulators. However these latter atoms are heavier, their crystal unit cells are larger and the characteristic energies smaller : thus $E_g = 1.1$ eV in silicon and 0.75 eV in germanium.

Remember that a *full band has no conduction*, the application of an electric field producing no change of the macroscopic occupation of the levels. This is equally valid for the $1s$ band of the lithium metal and at zero temperature for the valence band of insulators like Si or Ge.

8.4.2   Variation of Metal Resistance versus Temperature

We now are able to justify the methods and results on metals of the previous chapter, by reference to the case of lithium discussed above : their bands consist in one or several full bands, inactive under the application of an electric field, and a conduction band partly filled at zero temperature. The density of states at the bottom of this latter band is analogous to that of an electron in a box, under the condition that the free electron mass is replaced by the effective mass at the bottom of the conduction band. Indeed we already anticipated this result when we analyzed the specific heats of copper and silver (7.2.2).

The electrical conductivity σ is related to the linear effect of an applied field $\vec{E}$ on the conduction electrons. An elementary argument allows one to establish its expression $\sigma = ne^2\tau/m^*$, where n is the electron concentration in the conduction band, e the electron charge, τ the mean time between two collisions and m^* the effective mass. A more rigorous analysis, in the framework of the transport theory, justifies this formula which is far from being obvious : the only excitable electrons are in the vicinity of the Fermi level, their velocity in the absence of electric field is v_F.

The phenomenological τ term takes into account the various mechanisms which limit the conductivity : in fact the probabilities of these different independent mechanisms are additive, so that

$$\frac{1}{\tau} = \frac{1}{\tau_{\text{vibr}}} + \frac{1}{\tau_{\text{imp}}} + \cdots \tag{8.37}$$

$1/\tau_{\text{vibr}}$ corresponds to the probability of diffusion by the lattice vibrations, which increases almost linearly with the temperature T, $1/\tau_{\text{imp}}$ to the diffusion by neutral or ionized impurities, a mechanism present even at zero temperature.

Then

$$\frac{1}{\sigma} = \rho = \frac{m^*}{ne^2} \left(\frac{1}{\tau_{\text{vibr}}} + \frac{1}{\tau_{\text{imp}}} + \cdots \right) \tag{8.38}$$

$$\rho = \rho_{\text{vibr}}(T) + \rho_{\text{imp}} + \cdots$$

Thus the resistivity terms arising from the various mechanisms add up, this is the empirical Matthiessen law, stated in the middle of the 19th century.

Consequently, the *resistance of a metal increases with temperature* according to a law practically linear in T, the dominating mechanism being the electron's diffusion on the lattice vibrations. This justifies the use of the platinum resistance thermometer as a secondary temperature standard between $-183°$ C and $+630°$ C.

8.4.3 Variation of Insulator's Conductivity Versus Temperature; Semiconductors

a) Concentrations of Mobile Carriers in Equilibrium at Temperature T

We have just seen that at zero temperature an electric field has no effect on an insulator, since its valence band is totally full and its conduction band empty. However there are processes to excite such systems. The two main ones are :

- an optical excitation : a photon of energy $h\nu > E_g$ will promote an electron from the valence band into the conduction band. This is the absorption phenomenon, partly responsible for the color of these solids. It creates a nonequilibrium thermal situation.

- a thermal excitation : the thermal motion is also at the origin of transitions from the valence band to the conduction band. We are now going to detail this latter phenomenon, which produces a thermal equilibrium at temperature T.

When an electron is excited from the valence band, this band then makes a transition from the state (full band) to the state (full band minus one electron), which allows one to define a quasi-particle, the "hole," which is the lack of an electron. We already noted in chapter 7.1.1 the symmetry, due to the shape of the Fermi distribution, between the occupation factor $f(\varepsilon)$ of energy states above μ, and the probability $1 - f(\varepsilon)$ of the states below μ for being empty. The following calculation will show the similarity between the electron or hole role.

The electrons inside the insulator follow the Fermi-Dirac statistics and at *zero temperature* the conduction band is empty and the valence band full. This means that ε_F lies between ε_v, the top of the valence band and ε_c, the minimum of the conduction band. At weak temperature T ($k_B T \ll E_g$) one expects that the chemical potential μ still lies inside the energy band gap.

We are going to estimate $n(T)$, the *number of electrons* present at T in the *conduction band* and $p(T)$, the *number of electrons missing* (or holes present) in the *valence band*, and to deduce the energy location $\mu(T)$ of the chemical potential. These very few carriers will be able to react to an applied electric field $\vec{E}$: indeed in the conduction band the electrons are very scarce at usual temperatures and they are able to find vacant states in their close neighborhood when excited by $\vec{E}$; in the same way the presence of several holes in the valence band permits the valence electrons to react to $\vec{E}$.

The volume concentration $n(T)$ of *electrons* present in the *conduction band* at temperature T in the solid of volume Ω is such that

$$\Omega n(T) = \int_{\varepsilon_c}^{\varepsilon_{c\max}} D(\varepsilon) f(\varepsilon) d\varepsilon = \int_{\varepsilon_c}^{\varepsilon_{c\max}} D(\varepsilon) \frac{1}{\exp \beta(\varepsilon - \mu) + 1} d\varepsilon \qquad (8.39)$$

where ε_c is the energy of the conduction band minimum, $\varepsilon_{c\max}$ the energy of its maximum.

One assumes, and it will be verified at the end of the calculation, that μ is distant of this conduction band of several $k_B T$, which leads to $\beta(\varepsilon_c - \mu) \gg 1$, so that one can approximate $f(\varepsilon)$ by a Boltzmann factor. Then

$$\Omega n(T) \simeq \int_{\varepsilon_c}^{\varepsilon_{c\max}} D(\varepsilon) e^{-\beta(\varepsilon - \mu)} d\varepsilon \qquad (8.40)$$

Because of the very fast exponential variation, the integrand has a significant value only if $\varepsilon - \varepsilon_c$ does not exceeds a few $k_B T$. Consequently, one can extend the upper limit of the integral to $+\infty$, as the added contributions are absolutely negligible. Moreover, only the expression of $D(\varepsilon)$ at the bottom of the conduction band really matters : we saw in (8.34) that an effective mass, here m_c, can be introduced, which expresses the band curvature in the vicinity of

$\varepsilon = \varepsilon_c$. Whence

$$n(T) \simeq \int_{\varepsilon_c}^{\infty} \frac{1}{2\pi^2\hbar^3} (2m_c)^{3/2} \sqrt{\varepsilon - \varepsilon_c}\, e^{-\beta(\varepsilon-\mu)}\, d\varepsilon \qquad (8.41)$$

$$= \frac{1}{2\pi^2\hbar^3} e^{-\beta(\varepsilon_c-\mu)} (2m_c kT)^{3/2} \int_0^{\infty} \sqrt{x}\, e^{-x}\, dx$$

$$= 2\left(\frac{2\pi m_c k_B T}{h^2}\right)^{3/2} \exp\left(-\frac{\varepsilon_c - \mu}{k_B T}\right) \qquad (8.42)$$

In the same way one calculates the *hole concentration* $p(T)$ in the *valence band*, the absence of electrons at temperature T being proportional to the factor $1 - f(\varepsilon)$:

$$\Omega p(T) = \int_{\varepsilon_{v\min}}^{\varepsilon_v} D(\varepsilon)[1 - f(\varepsilon)]\, d\varepsilon \qquad (8.43)$$

One assumes that μ is at a distance of several $k_B T$ from ε_v, the maximum of the valence band, of minimum $\varepsilon_{v\min}$. The neighborhood of ε_v is the only region that will intervene in the integral calculation because of the fast decrease of $1 - f(\varepsilon) \simeq e^{-\beta(\mu-\varepsilon)}$. In this region $D(\varepsilon)$ is parabolic and is approximated by

$$D(\varepsilon) \simeq \frac{\Omega}{2\pi^2\hbar^3} (2m_v)^{3/2} \sqrt{\varepsilon_v - \varepsilon} \qquad (8.44)$$

where $m_v > 0$, the "hole effective mass," expresses the curvature of $D(\varepsilon)$ in the neighborhood of the valence band maximum. One gets :

$$p(T) = 2\left(\frac{2\pi m_v k_B T}{h^2}\right)^{3/2} \exp\left(-\frac{\mu - \varepsilon_v}{k_B T}\right) \qquad (8.45)$$

In the calculations between formulae (8.40) and (8.45) we did not make any assumption about the way $n(T)$ and $p(T)$ are produced. These expressions are general, as is the product of (8.42) by (8.45) :

$$n(T) \cdot p(T) = 4\left(\frac{2\pi k_B T}{h^2}\right)^3 (m_c m_v)^{3/2} \exp\left(-\frac{E_g}{k_B T}\right) \qquad (8.46)$$

with $E_g = \varepsilon_c - \varepsilon_v$.

b) Pure (or Intrinsic) Semiconductor

In the case of a thermal excitation, for each electron excited into the conduction band a hole is left in the valence band. One deduces from (8.46) :

$$n(T) = p(T) = n_i(T) = 2\left(\frac{2\pi k_B T}{h^2}\right)^{3/2} (m_c m_v)^{3/4} \exp\left(-\frac{E_g}{2k_B T}\right) \qquad (8.47)$$

For this *intrinsic* mechanism, the variation of $n_i(T)$ versus temperature is dominated by the $\exp(-E_g/2k_BT)$ term. In fact, here the thermal mechanism creates two charge carriers with a threshold energy E_g.

The very fast variation of the exponential is responsible for the distinction between *insulators* and *semiconductors*. Indeed at room temperature $T = 300$ K, which corresponds to approximately $\frac{1}{40}$ eV, let us compare the diamond situation where $E_g = 5.2$ eV $[\exp(-E_g/2k_BT) = \exp(-104) = 6.8 \times 10^{-46}]$ and that of silicon, where $E_g = 1.1$ eV $[\exp(-22) = 2.8 \times 10^{-10}]$. In 1 cm^3 of diamond, there is no electron in the conduction band, whereas $n_i(T)$ is equal to 1.6×10^{16} m^{-3} at 300 K in silicon.

A *semiconductor is an insulator with a "not too wide" band gap*, which allows the excitation of some electrons into the conduction band at room temperature. The most commonly used semiconductors, apart from silicon, are germanium and gallium arsenide GaAs $(E_g = 1.52$ eV$)$.

The chemical potential of an intrinsic semiconductor is determined by substituting expression (8.47) of $n_i(T)$ into (8.41). One obtains :

$$\mu = \frac{\varepsilon_c + \varepsilon_v}{2} + \frac{3}{4}k_BT \ln \frac{m_v}{m_c} \qquad (8.48)$$

The Fermi level ε_F, which is the chemical potential at $T = 0$ K, lies in the middle of the band gap and μ does not move much apart from this location when T increases $(k_BT \ll E_g)$. Consequently, the hypotheses that μ lies very far from the band edges as compared to k_BT, made when calculating $n(T)$ and $p(T)$, are valid.

Under the effect of an external electric field, both electrons and holes participate in the electrical current : indeed the current includes contributions from both the conduction and the valence bands. As will be studied in Solid State Physics courses, the holes behave like positive charges and the total electrical conductivity in the semiconductor is the sum of the valence band and conduction band currents :

$$\sigma = \sigma_n + \sigma_p$$
$$= \left(ne^2 \frac{\tau_n}{m_c} + pe^2 \frac{\tau_p}{m_v} \right) \qquad (8.49)$$

Here τ_n (respectively τ_p) is the electron (respectively hole) collision time.

The temperature variation of σ in an intrinsic semiconductor is dominated by the variation of $\exp(-E_g/2kT)$, the other factors varying like powers of T. Consequently, σ increases very fast with T, thus its reciprocal, the resistivity and the measured *resistance of a semiconductor decrease with temperature* because the number of carriers increases.

c) Doped Semiconductor

The density of intrinsic carriers at room temperature is very weak and practically impossible to attain : for instance in silicon about 1 electron out of 10^{12} is in the conduction band due to thermal excitation. To achieve the intrinsic concentration the studied sample should contain neither impurities nor defects ! It is more realistic to control the conductivity by selected impurities, the *dopants*.

Thus phosphorus atoms can be introduced into a silicon crystal : phosphorus is an element of column V of the periodic table, which substitutes for silicon atoms in some lattice sites. Each phosphorus atom brings one more electron than required by the tetrahedral bonds of Si. At low temperature this electron remains localized near the phosphorus nucleus, on a level located 44 meV below the bottom of the conduction band. When the temperature increases, the electron is ionized into the conduction band. Phosphorus is a *donor*.

The concentration of conduction electrons will be determined by the concentration N_D in P atoms, of typical value $N_D \sim 10^{22}\,\mathrm{m}^{-3}$, chosen much larger than the intrinsic and the residual impurity concentrations, but much smaller than the Si atoms concentration.

In such a material, because of (8.46), the hole concentration is very small with respect to n_i, the current is carried by the electrons, the material is called of *n-type*.

In a symmetrical way, the substitution in the silicon lattice of a Si atom by a boron atom creates an *acceptor state* : boron, an element of column III of the periodic table, only has three peripheral electrons and thus must capture one extra electron in the valence band to satisfy the four tetrahedral bonds of the crystal. This is equivalent to creating a hole, which, at low temperature, is localized 46 meV above the silicon valence band and at higher temperature is delocalized inside the valence band.

In boron-doped silicon, the conductivity arises from the holes, the semiconductor is of *p-type*.

In Solid State Physics courses, the statistics of doped semiconductors is studied in more detail : one finds in particular that the chemical potential μ moves versus temperature inside the band gap (in a n-type material, at low temperature it is close ε_c, in the vicinity of the donors levels and returns to the middle of the band gap, like in an intrinsic material, at very high temperature). The semiconductors constituting the electronic compounds are generally doped.

8.5   Examples of Semiconductor Devices

Here we only give a flavor of the variety of configurations and uses. In specialized Solid State Physics courses, one will describe in more details the semiconductors properties. The aim here is to show that, just using the basic elements of the present chapter, one can understand the principle of operation of three widespread devices.

8.5.1   The Photocopier : Photoconductivity Properties

Illuminating a semiconductor by light of energy larger than E_g promotes an electron into the conduction band and leaves a hole in the valence band. The conductivity increase is $\Delta\sigma$, proportional to the light flux. In the xerography process, the photocopier drum is made of selenium which, under the illumination effect, shifts from a very insulator state to a conducting state. This property is utilized to fix the pigments (the *toner* black powder) which are next transferred to the paper and then cooked in an oven to be stabilized on the paper sheet.

8.5.2   The Solar Cell : an Illuminated $p - n$ Junction

When, in the same solid, a p region is placed next to a n region, the electrons of the n region diffuse toward the holes of the p region : some of these mobile charges neutralize and a so-called *space charge region* becomes emptied of mobile carriers. The dopants still remain in the region, their charged nuclei create an electric field which prevents any additional diffusion. This effect can also be interpreted from the requirement of equal chemical potentials in the n and p region at equilibrium : in the neighborhood (a few μm) of the region of doping change, there prevails an intense internal electric field $\vec{E}_i$. If the sun is illuminating this region, the excited electron and hole, of opposite charges, are separated by $\vec{E}_i$. An external use circuit will take advantage of the current produced by this light excitation. We will reconsider these devices, called "solar cells," at the end of next chapter when we will deal with thermal radiation.

8.5.3 Compact Disk Readers : Semiconductor Quantum Wells

It was seen in § 8.3.3 that starting from the energy levels of the constituting atoms one gets the energy bands of a solid. Their location depends of the initial atomic levels, thus of the chemical nature of the material. For almost thirty years it has been possible, in particular using Molecular Beam Epitaxy (MBE), to sequentially crystallize different semiconductors, which have the same crystal lattice but different chemical composition : a common example is GaAs and $Ga_{1-x}Al_xAs$.

In such semiconductor sandwiches the bottom of the conduction band (or the top of the valence band) does not lie at the same energy in GaAs and in $Ga_{1-x}Al_xAs$. One thus tailors *quantum wells*, to fashion. Their width usually ranges from 5 to 20 nm, their energy depth is of a few hundreds of meV, and the energies of their localized levels is determined by their "geometry," as taught in Quantum Mechanics courses.

These quantum wells are much used in optoelectronics, for example as diodes or lasers emitters, in the telecommunication networks by optical fibers, as readers of compact disks, of code-bars in supermarkets, etc.

Summary of Chapter 8

One has first to study the quantum states of a periodic solid. By analogy with the problem of the double potential well of the Quantum Mechanics course, one considers an electron on an infinite periodic chain, in a model of independent electrons, of the Linear Combination of Atomic Orbitals type. The hamiltonian eigenstates are Bloch states, the eigenenergies are regrouped into allowed bands separated by forbidden bands (or band gaps). The important parameter is the coupling term between neighboring sites.

In a perfect crystal, the velocity of an electron would be maintained forever.

The finite macroscopic dimension of a crystal allows one to define a density of states and to show that the number of orbital states is the product of the number of cells in the crystal by the number of orbital states corresponding to each initial atomic state. Filling these states together with considering the spin properties, allows one to distinguish between insulators and metals.

We calculated the number of mobile carriers of an insulator versus temperature. The resistance of an insulator exponentially decreases when T increases, the resistance of a metal increases almost linearly with T.

A semiconductor is an insulator with a "rather small" band gap energy.

Chapter 9

Bosons : Helium 4, Photons, Thermal Radiation

We learnt in chapter 5 that bosons are particles of integer or null spin, with a symmetrical wave function, i.e. which is unmodified in the exchange of two particles. Bosons can be found in arbitrary number on a one-particle state of given energy ε_k. We are going now to discuss in more detail the properties of systems following the Bose-Einstein statistics. We will begin with the case of material particles and take the example of ^4_2He, the helium isotope made up of two protons and two neutrons. The Bose-Einstein condensation occurring in this case is related to the helium property of superfluidity. Then we will study the statistics in equilibrium of photons, relativistic particles of zero mass. The photon statistics has an essential importance for our life on the earth through the greenhouse effect. Moreover, it played a major conceptual part : indeed it is in order to interpret the experimental thermal radiation behavior that Max Planck (1858-1947) produced the quanta hypothesis (1901). Finally we will take a macroscopic point of view and stress the importance of thermal radiation phenomena in our daily experience.

9.1 Material Particles

The ability of bosons of given non-zero mass to gather on the same level under a low enough temperature yields to a phase transition, the Bose-Einstein condensation.

9.1.1   Thermodynamics of the Boson Gas

Let us recall the average occupation number $\langle n_k \rangle$ [Eq.(6.26)] by bosons of chemical potential μ, of the energy level ε_k at temperature $T = \dfrac{1}{k_B \beta}$:

$$\langle n_k \rangle = \frac{1}{e^{\beta(\varepsilon_k - \mu)} - 1} \tag{9.1}$$

As n_k is positive or zero, the denominator exponent should be positive for any ε_k, which implies that the chemical potential μ must be smaller than any ε_k, including ε_1, the energy of the fundamental state. At fixed T and μ, $\langle n_k \rangle$ decreases when ε_k increases. The μ value is fixed by the condition

$$\sum_k \langle n_k \rangle = N \tag{9.2}$$

where N is the total particles number of the considered physical system.

From $\langle n_k \rangle$ one can calculate all the thermodynamical parameters and in particular the average energy :

$$U = \sum_k \varepsilon_k \langle n_k \rangle \tag{9.3}$$

the grand potential A and its partial derivatives S and P

$$A = -k_B T \sum_k \ln(1 + \langle n_k \rangle) \tag{9.4}$$

$$dA = -Pd\Omega - SdT - Nd\mu \tag{9.5}$$

In the large volume limit, (9.1) is replaced by the Bose-Einstein distribution, a continuous function given by

$$f(\varepsilon) = \frac{1}{e^{\beta(\varepsilon - \mu)} - 1} \tag{9.6}$$

The physical parameters are calculated in this case using the density of states $D(\varepsilon)$, then expressing the constraint replacing (9.2) on the particles number and verifying that $\mu < \varepsilon_1$. Here (§ 9.1) one is concerned by *free bosons*, with no mutual interaction, therefore their energy is reduced to $\varepsilon = \dfrac{p^2}{2m}$ (apart from the "box" potential producing the quantization, leading to the expression of $D(\varepsilon)$), so that $\varepsilon_1 = 0$ and $\mu < 0$.

Then for particles with a spin s, in the three-dimensional space, one gets, taking $D(\varepsilon) = C\Omega\sqrt{\varepsilon}$, where $C = \dfrac{2s+1}{4\pi^2}\left(\dfrac{2m}{\hbar^2}\right)^{3/2}$ [cf. Eq. (6.64)],

$$N = \int_0^\infty f(\varepsilon)D(\varepsilon)d\varepsilon = C\Omega\int_0^\infty \frac{\sqrt{\varepsilon}d\varepsilon}{e^{\beta(\varepsilon-\mu)}-1} \tag{9.7}$$

$$U = \int_0^\infty \varepsilon f(\varepsilon)D(\varepsilon)d\varepsilon = C\Omega\int_0^\infty \frac{\varepsilon^{3/2}d\varepsilon}{e^{\beta(\varepsilon-\mu)}-1} \tag{9.8}$$

$$A = -k_BT\int_0^\infty \ln[1+f(\varepsilon)]D(\varepsilon)d\varepsilon$$

$$A = C\Omega k_BT\int_0^\infty \sqrt{\varepsilon}\ln[1-e^{-\beta(\varepsilon-\mu)}]d\varepsilon \tag{9.9}$$

$$= -\frac{2}{3}C\Omega\int_0^\infty \frac{\varepsilon^{3/2}d\varepsilon}{e^{\beta(\varepsilon-\mu)}-1} \tag{9.10}$$

Integrating (9.9) by parts, one finds (9.10) which is equal to (9.8) to the factor $-\dfrac{2}{3}$, so that

$$A = -\frac{2}{3}U = -P\Omega \tag{9.11}$$

This is the state equation of the bosons gas, that is,

$$P = \frac{2}{3}\frac{U}{\Omega} \tag{9.12}$$

which takes the same form as for the fermions (7.13), although the physical properties are totally different.

9.1.2 Bose-Einstein Condensation

The above expressions are only valid when the condition $\mu < 0$ is fulfilled (see above). Now when T decreases at constant density N/Ω, since β increases the differences $(\varepsilon - \mu)$ must decrease in order that N/Ω keeps its value (9.7). This means that μ gets closer to zero.

Now consider the *limit situation* $\mu = 0$. Relation (9.7) then becomes

$$N = C\Omega\int_0^\infty \frac{\sqrt{\varepsilon}d\varepsilon}{e^{\beta\varepsilon}-1} \tag{9.13}$$

$$N = \frac{C\Omega}{\beta_B^{3/2}}\int_0^\infty \frac{\sqrt{x}dx}{e^x-1} = \frac{C\Omega}{\beta_B^{3/2}}I\left(\frac{1}{2}\right) \tag{9.14}$$

i.e., using the mathematical tables of the present book,

$$N = \frac{C\Omega}{\beta_B^{3/2}}\cdot 2,315 \tag{9.15}$$

Relation (9.15) defines a specific temperature $T_B = 1/k_B\beta_B$ which is only related to the density $n = N/\Omega$:

$$T_B = \frac{1}{k_B}\left(\frac{N}{C\Omega}\frac{1}{2,315}\right)^{2/3} \tag{9.16}$$

$$= \frac{6,63\hbar^2}{2mk_B}\frac{n^{2/3}}{(2s+1)^{2/3}} \tag{9.17}$$

In (9.17) C was replaced by its expression $C = \dfrac{2s+1}{4\pi^2}\left(\dfrac{2m}{\hbar^2}\right)^{3/2}$, in which the spin of the $_2^4$He particles is zero.

The temperature T_B, or Bose temperature, would correspond in the above calculation to $\mu = 0$, whereas the situation of § 9.1.1 is related to $\mu < 0$. Since the integrals in (9.7) and (9.13) are both equal to $N/C\Omega$, to maintain this value constant it is necessary that β (corresponding to $\mu < 0$) be smaller than β_B (associated to $\mu = 0$). *Consequently, (9.7) is only valid for $T > T_B$.* When lowering the temperature to T_B, μ tends to zero, thus getting closer to the fundamental state of energy $\varepsilon_1 = 0$.

For $T < T_B$ *the analysis has to be reconsidered : the population of level ε_1 has to be studied separately*, as it would diverge if one were using expression (9.1) without caution! Now the population N_1 of ε_1 is at most equal to N. Expression (9.1) remains valid but μ cannot be strictly zero.

Assume that N_1 is a macroscopic number of the order of N :

$$N_1 = \frac{1}{e^{\beta(\varepsilon_1-\mu)}-1} = \frac{1}{e^{-\beta\mu}-1} \tag{9.18}$$

The corresponding value of μ is given by :

$$e^{-\beta\mu} = 1 + \frac{1}{N_1}$$
$$\mu \simeq -\frac{k_B T}{N_1} \tag{9.19}$$

μ is very close to zero because it is the ratio of a microscopic energy to a macroscopic number of particles. The first excited level ε_2 has the energy $\dfrac{\hbar^2}{2mL^2} \propto \Omega^{-2/3}$, whereas the chemical potential is in Ω^{-1}, so that μ is very small as compared to ε_2.

The $N - N_1$ particles which are not located in ε_1 are distributed among the excited states, for which one can take $\mu = 0$ without introducing a large error, and adapt the condition (9.7) on the particle number

$$N - N_1 = C\Omega \int_{\varepsilon_2}^{\infty} \frac{\sqrt{\varepsilon}\, d\varepsilon}{e^{\beta\varepsilon} - 1} \qquad (9.20)$$

$$\simeq C\Omega \int_{\varepsilon_1=0}^{\infty} \frac{\sqrt{\varepsilon}\, d\varepsilon}{e^{\beta\varepsilon} - 1} \qquad (9.21)$$

Indeed shifting the limit of the integral from ε_2 to ε_1 is adding a relative contribution to (9.20) of the order of $(\varepsilon_2/k_B T)^{1/2}$. Since the volume Ω is macroscopic, this quantity is very small, ε_2 being much less than $k_B T$.

One recognizes in (9.21) the integral (9.13) under the condition of replacing β_B by β. Whence :

$$\frac{N}{T_B^{3/2}} = \frac{N - N_1}{T^{3/2}} \qquad (9.22)$$

$$N_1 = N \left[1 - \left(\frac{T}{T_B} \right)^{3/2} \right] \qquad (9.23)$$

Thus, when lowering the temperature, different regimes take place :

- as long as $T > T_B$, the chemical potential is negative and the particles are distributed among all the microscopic states of the system according to the Bose-Einstein distribution (9.6) ;
- when $T = T_B$, μ almost vanishes (it is very small but negative) ;
- when $T < T_B$, the particles number N_1 on the fundamental level becomes macroscopic, this is the "Bose condensation." One thus realizes a macroscopic quantum state. The distribution of particles on the excited levels corresponds to $\mu \simeq 0$. When T continues to decrease, N_1 increases and tends to N when T tends to zero ; μ is practically zero.

Going through T_B corresponds to a *phase transition* : for $T < T_B$ the Bose gas is degenerate, which means that a macroscopic particles number is located in the same quantum state. This corresponds for $_2^4$He to the *superfluid* state. The phenomenon of helium superfluidity is very complex and the interactions between atoms, neglected up to now, play a major part in this phase. The above calculation predicts a transition to the superfluid state at 3.2 K in standard conditions (^{4}He specific mass $=140$ kg/m^3 under the atmospheric pressure) whereas the transition occurs at $T_\lambda = 2.17$ K. This is a second-order phase transition (the free energy is then finite, continuous and derivable ; the specific heat is discontinuous).

The *superconductivity* of some solids also corresponds to a Bose condensation (2003 Nobel prize of A. Abrikosov and V. Ginzburg). It was possible to observe the *atoms' Bose condensation* by trapping and cooling atomic beams at temperatures of the order of a mK in "optical molasses" (1997 Nobel prices of S. Chu, C. Cohen-Tannoudji and W.D. Phillips).

9.2   Bose-Einstein Distribution of Photons

Our daily experience teaches us that a heated body radiates, as is the case for a radiator heating a room or an incandescent lamp illuminating us through its visible radiation. Understanding these mechanisms, a challenge at the end of the 19th century, was at the origin of M. Planck's quanta hypothesis (1901). For more clarity we will not follow the historical approach but rather use the tools provided by this course. Once the results obtained, we will return to the problems Planck was facing.

9.2.1   Description of the Thermal Radiation ; the Photons

Consider a cavity of volume Ω heated at temperature T (for example an oven) and previously evacuated. The material walls of this cavity are made of atoms, the electrons of which are promoted into excited states by the energy from the heat source. During their deexcitation, these electrons radiate energy, under the form of an electromagnetic field which propagates inside the cavity, is absorbed by other electrons associated to other atoms, and so forth, and this process leads to a thermal equilibrium between the walls and the radiation enclosed inside the cavity.

If the cavity walls are perfect reflectors, classical electromagnetism teaches us that the wave electrical field $\vec{E}$ is zero inside the perfect conductor and normal to the wall on the vacuum side (due to the condition that $\vec{E}_{\text{tangential}}$ is continuous), and that the corresponding magnetic field $\vec{B}$ must be tangential. In order to produce the resonance of this cavity and to establish *stationary waves*, the cavity dimension L should contain an integer number of half-wavelengths λ. The wave vector $k = 2\pi/\lambda$ and L are then linked.

An electromagnetic wave, given in complex notation by $\vec{E}(\omega, \vec{k}) \exp i(\vec{k} \cdot \vec{r} - \omega t)$ is characterized by its wave vector $\vec{k}$ and its frequency ω ; it can have two independent polarization states (either linear polarizations along two perpendicular directions, or left or right circular polarizations) on which the direction of the electric field in the plane normal to $\vec{k}$ is projected. Finally its intensity is proportional to $|\vec{E}(\omega, \vec{k})|^2$.

Since the interpretation of the photoelectric effect by A. Einstein in 1905, it has been known that the light can be described both in terms of waves and of particles, which are the photons. The quantum description of the electromagnetic field is taught in advanced Quantum Mechanics courses. Here we only specify that the photon is a relativistic particle of zero mass, so that the general relativistic relation between the energy ε and the momentum p, $\varepsilon = \sqrt{p^2 c^2 + m_0^2 c^4}$, where c is the light velocity in vacuum, reduces in this case to $\varepsilon = pc$. Although the photon spin is equal to 1, due to its zero mass it only has two distinct spin states. The set of data consisting in $\vec{p}$ and the photon spin value defines a mode.

The wave parameters $\vec{k}$ and ω and the photon parameters are related through the Planck constant, as summarized in this table :

electromagnetic wave	**particle :** photon
wave vector $\vec{k}$	$\vec{p} = \hbar\vec{k}$ momentum
frequency ω	$\varepsilon = \hbar\omega$ energy
intensity $|\vec{E}(\vec{k},\omega)|^2$	number of photons
polarization (2 states)	spin (2 values)

Let us return to the cavity in thermal equilibrium with the radiation it contains : its dimensions, in $\Omega^{1/3}$, are very large with respect to the wavelength of the considered radiation. The exact surface conditions do not matter much : as in the particles case (see § 6.4.1b)), rather than using the stationary wave conditions, we prefer the *periodic limit conditions* ("Born-Von Kármán" conditions) : in a thought experiment we close the system on itself, which quantizes $\vec{k}$ and consequently, $\vec{p}$ and the energy. Assuming the cavity is a box of dimensions L_x, L_y, L_z, one obtains :

$$\vec{k} = (k_x, k_y, k_z) = 2\pi \left(\frac{n_x}{L_x}, \frac{n_y}{L_y}, \frac{n_z}{L_z} \right) \tag{9.24}$$

where n_x, n_y, n_z are positive or negative integers, or zero.

Then

$$\vec{p} = h \left(\frac{n_x}{L_x}, \frac{n_y}{L_y}, \frac{n_z}{L_z} \right) \tag{9.25}$$

$$\varepsilon = pc = \hbar\omega = h\nu \tag{9.26}$$

9.2.2 Statistics of Photons, Bosons in Non-Conserved Number

In the photon emission and absorption processes of the wall atoms, the number of photons varies, and there are more photons when the walls are hotter : in this system the constraint on the conservation of particles number, that existed in § 9.1, is lifted. This constraint was defining the chemical potential μ as a Lagrange multiplier, which no longer appears in the probability law for photons, i.e., the chemical potential is zero for photons.

Let us consider the special case where only a single photon mode is possible, i.e., a single value of $\vec{p}$, of the polarization and of ε in the cavity at thermal equilibrium at temperature T. Any number of photons is possible, thus the partition function $z(\varepsilon)$ is equal to

$$z(\varepsilon) = \sum_{n=0}^{\infty} e^{-\beta n \varepsilon} = \frac{1}{1 - e^{-\beta \varepsilon}} \tag{9.27}$$

and the average energy at T of the photons in this mode is given by

$$\langle n \rangle \varepsilon = -\frac{1}{z(\varepsilon)} \frac{\partial z(\varepsilon)}{\partial \beta} = \frac{\varepsilon}{e^{\beta \varepsilon} - 1}$$

For the occupation factor of this mode, one finds again the Bose-Einstein distribution in which $\mu = 0$, i.e.

$$f(\varepsilon) = \frac{1}{e^{\beta \varepsilon} - 1} \tag{9.28}$$

[Note that in the study of the specific heat of solids (see §2.4.5) a factor similar to (9.28) appeared : in the Einstein and Debye models the quantized solid vibrations are described by harmonic oscillators. Changing the vibration state of the solid from $\left(n + \frac{1}{2}\right) \hbar \omega$ to $\left(n + \frac{1}{2} + 1\right) \hbar \omega$ is equivalent to creating a quasi-particle, called a *phonon*, which follows the Bose-Einstein statistics. The number of phonons is not conserved].

If now one accounts for all the photons modes, for a *macroscopic* cavity volume Ω, one can define the densities of states $D_{\vec{k}}(\vec{k}), D_{\vec{p}}(\vec{p}), D(\varepsilon)$ such that the number dn of modes with a wave vector between $\vec{k}$ and $\vec{k} + d\vec{k}$ is equal to dn :

$$dn = 2 \cdot \frac{\Omega}{(2\pi)^3} d^3\vec{k} = D_{\vec{k}}(\vec{k}) d^3\vec{k}$$

$$dn = 2 \cdot \frac{\Omega}{h^3} d^3\vec{p} = D_{\vec{p}}(\vec{p}) d^3\vec{p}$$

that is,

$$D_{\vec{k}}(\vec{k}) = \frac{\Omega}{4\pi^3} \quad , \quad D_{\vec{p}}(\vec{p}) = \frac{2\Omega}{h^3} \tag{9.29}$$

The factor 2 expresses the two possible values of the spin of a photon of fixed $\vec{k}$ (or $\vec{p}$).

Since the energy ε only depends of the modulus $\vec{p}$, one obtains $D(\varepsilon)$ from

$$D(\varepsilon) d\varepsilon = 2 \frac{\Omega}{h^3} 4\pi p^2 dp = 8\pi \frac{\Omega}{h^3 c^3} \varepsilon^2 d\varepsilon$$

$$D(\varepsilon) = \frac{8\pi\Omega}{h^3 c^3} \varepsilon^2 \tag{9.30}$$

In this "large volume limit", one can express the thermodynamical parameters.

The thermodynamical potential to be considered here is the free energy F (since $\mu = 0$)

$$F = -k_B T \ln Z = k_B T \int_0^\infty \ln(1 - e^{-\beta\varepsilon}) D(\varepsilon) d\varepsilon \tag{9.31}$$

The internal energy is written

$$U = \int_0^\infty \varepsilon D(\varepsilon) f(\varepsilon) d\varepsilon$$

$$= \int_0^\infty \frac{8\pi\Omega}{h^3 c^3} \frac{\varepsilon^3}{e^{\beta\varepsilon} - 1} d\varepsilon = \int_0^\infty \frac{8\pi\Omega}{c^3} \frac{h\nu^3 d\nu}{e^{\beta h\nu} - 1} \tag{9.32}$$

i.e.,

$$U = \int_0^\infty \Omega u(\nu) d\nu$$

with

$$u(\nu) = \frac{8\pi h}{c^3} \frac{\nu^3}{e^{\beta h\nu} - 1} \tag{9.33}$$

The *photon spectral density in energy* $u(\nu)$ is defined from the energy contribution dU of the photons present inside the volume Ω, with a frequency between ν and $\nu + d\nu$:

$$dU = \Omega u(\nu) d\nu \tag{9.34}$$

The expression (9.33) of $u(\nu)$ is called the *Planck law*.

The total energy for the whole spectrum is given by

$$U = \frac{8\pi\Omega}{h^3 c^3} (k_B T)^4 \int_0^\infty \frac{x^3 dx}{e^x - 1} \tag{9.35}$$

where the dimensionless integral is equal to $\Gamma(4)\zeta(4)$ with $\Gamma(4) = 3!$ and $\zeta(4) = \dfrac{\pi^4}{90}$ (see the section "Some useful formulae" in this book). Introducing $\hbar = h/2\pi$, the total energy becomes

$$U = \frac{\pi^2}{15} \Omega \frac{(k_B T)^4}{(\hbar c)^3} \tag{9.36}$$

The free energy U and the internal energy F are easily related : in the integration by parts of (9.31), the derivation of the logarithm introduces the Bose-Einstein distribution, the integration of $D(\varepsilon)$ provides a term in $\varepsilon^3/3 = \varepsilon.\varepsilon^2/3$. This exactly gives

$$F = -\frac{U}{3} = -\frac{\pi^2}{45}\Omega\frac{(k_BT)^4}{(\hbar c)^3}$$

[to be compared to (9.11), which is valid for material particles with a density of states in $\varepsilon^{1/2}$, whereas the photons density varies in ε^2]. Since

$$dF = -SdT - Pd\Omega \qquad (9.37)$$

one deduces

$$S = -\left(\frac{\partial F}{\partial T}\right)_\Omega = \frac{4\pi^2}{45}k_B\Omega\left(\frac{k_BT}{\hbar c}\right)^3 \qquad (9.38)$$

$$P = -\left(\frac{\partial F}{\partial \Omega}\right)_T = -\frac{F}{\Omega} \qquad (9.39)$$

$$P = \frac{\pi^2}{45}\frac{(k_BT)^4}{(\hbar c)^3} \qquad (9.40)$$

The pressure P created by the photons is called the *radiation pressure*.

9.2.3   Black Body Definition and Spectrum

The results obtained above are valid for a closed cavity, in which a measurement is strictly impossible because there is no access for a sensor. One is easily convinced that drilling a small hole into the cavity will not perturb the photons distribution but will permit measurements of the enclosed thermal radiation, through the observation of the radiation emitted by the hole. Besides any radiation coming from outside and passing through this hole will be trapped inside the cavity and will only get out after thermalization. This system is thus a perfect absorber, whence its name of *black body*.

In addition we considered that thermal equilibrium at temperature T is reached between the photons and the cavity. In such a system, at steady-state, for each frequency interval $d\nu$, the *energy absorbed by the walls exactly balances their emitted energy*, which is found in the photon gas in this frequency range, that is, $\Omega u(\nu)d\nu$. Therefore the parameters we obtained previously for the photon gas are also those characterizing the *thermal emission from matter at temperature T*, whether this emission takes place in a cavity or not, whether this matter is in contact with vacuum or not. We will then be able to compare the laws already stated, and in the first place the Planck law, to experiments performed on thermal radiation.

The expression (9.33) of $u(\nu)$ provides the spectral distribution of the thermal radiation versus the parameter β, i.e. versus the temperature T. For any T, $u(\nu)$ vanishes for $\nu = 0$ and tends to zero when ν tends to infinity. The variable is in fact $\beta h \nu$ and one finds that the maximum of $u(\nu)$ is reached for $\beta h \nu_{\text{max}} = 2.82$, that is,

$$\nu_{\text{max}} = 2.82 \frac{k_B T}{h} \qquad (9.41)$$

Relation (9.41) constitutes the Wien law.

When T increases, the frequency of the maximum increases and for $T_1 < T_2$ the whole curve $u(\nu)$ for T_2 is above that for T_1 (Fig. 9.1).

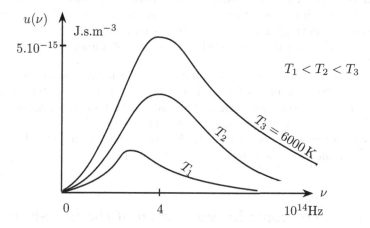

FIG. 9.1: Spectral density in energy of the thermal radiation for several temperatures.

$$\text{For small values of } \nu, \; u(\nu) \simeq k_B T \cdot \frac{8\pi}{c^3} \nu^2 \qquad (9.42)$$

$$\text{for } \nu \text{ large, } u(\nu) \simeq \frac{8\pi h}{c^3} \nu^3 e^{-\beta h \nu} \qquad (9.43)$$

At the end of the 19th century, photometric measures allowed one to obtain the $u(\nu)$ curves with a high accuracy. Using the then available Maxwell-Boltzmann statistics, Lord Rayleigh and James Jeans had predicted that the average at temperature T of the energy of oscillators should take the form (9.42), in $k_B T \nu^2$. If this latter law does describe the low-frequency behavior of the thermal radiation, it predicted an "ultraviolet catastrophe" [$u(\nu)$ would be an increasing function of ν, thus the UV and X emissions should be huge]. Besides, theoretical laws had been obtained by Wilhem Wien and experimentally confirmed by Friedrich Paschen : an exponential behavior at high frequency like in (9.43) and the law (9.41) of the displacement of the distribution maximum. It was the contribution of Max Planck in 1900 to guess the expression (9.33) of $u(\nu)$, assuming that the exchanges of energy between matter and radiation can only occur through discrete quantities, the quanta. Note that the ideas only slowly clarified until the advent in the mid 1920's of Quantum Mechanics, such as we know it now. It it remarkable that the paper by Bose, proposing the now called "Bose-Einstein statistics," preceded by a few months the formulation by Erwin Schroedinger of its equation!

The universe is immersed within an infrared radiation, studied in cosmology, the distribution of which very accurately follows a Planck law for $T = 2.7$ K. It is in fact a "fossil" radiation resulting from the cooling, by adiabatic expansion of the universe, of a thermal radiation at a temperature of 3000 K which was prevailing billions of years ago : according to expression (9.38) of the entropy, an adiabatic process maintains the product ΩT^3. The universe radius has thus increased of a factor 1000 during this period.

The sun is emitting toward us radiation with a maximum in the very near infrared, corresponding to T close to 6000 K. A "halogen" bulb lamp is emitting a radiation corresponding to the temperature of its tungsten filament (around 3000 K), which is immersed in a halogen gas to prevent its evaporation. We ourselves are emitting radiation with a maximum in the far infrared, corresponding to our temperature close to 300 K.

9.2.4   Microscopic Interpretation of the Bose-Einstein Distribution of Photons

Another schematic way to obtain the Bose-Einstein distribution, and consequently, the Planck law, consists in considering that the walls of the container are made of atoms with only two possible energy levels ε_0 and ε_1 separated by $\varepsilon_1 - \varepsilon_0 = h\nu$, and in assuming that the combined system (atoms-photons of energy $h\nu$) is at thermal equilibrium (Fig. 9.2). It is possible to rigorously calculate the transition probabilities between ε_0 and ε_1 (absorption), or between ε_1 and ε_0 (emission) under the effect of photons of energy $h\nu$. This is done in advanced Quantum Mechanics courses. Here we will limit ourselves to a more qualitative argument, admitting that in Quantum Mechanics a transition is expressed by a matrix element, which is the product of a term $\langle 1|V|0\rangle$ specific of the atoms with another term due to the photons. Besides, we will admit that an assembly of n photons of energy $h\nu$ is described by a harmonic oscillator, with energy levels splitting equal to $h\nu$, lying in the state $|n\rangle$.

The initial state contains N_0 atoms in the fundamental state, N_1 atoms in the state ε_1, and n photons of energy $h\nu$.

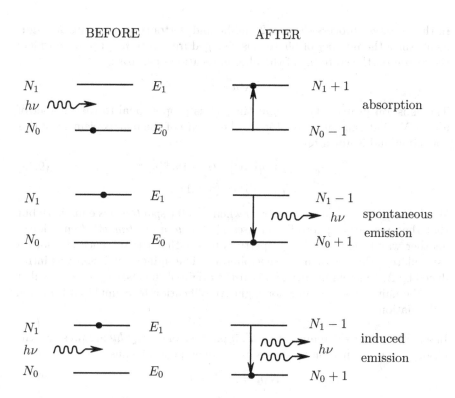

FIG. 9.2: Absorption, spontaneous and induced emission processes for an assembly of two-level atoms in thermal equilibrium with photons.

In the *absorption* process (Fig. 9.2 top), the number of atoms in the state of energy ε_0 becomes $N_0 - 1$, while N_1 changes to $N_1 + 1$; a photon is utilized, thus the photons assembly shifts from $|n\rangle$ to $|n - 1\rangle$. This is equivalent to considering the action on the state $|n\rangle$ of the photon annihilation operator, i.e., according to the Quantum Mechanics course,

$$a|n\rangle = \sqrt{n}\,|n - 1\rangle \tag{9.44}$$

The absorption probability per unit time P is proportional to the number of atoms N_0 capable of absorbing a photon, to the square of the matrix element appearing in the transition, and is equal, to a multiplicative constant, to :

$$P_a = N_0\,|\langle 1|V|0\rangle|^2\,|\langle n - 1|a|n\rangle|^2 \tag{9.45}$$

$$= N_0\,|\langle 1|V|0\rangle|^2\,n \tag{9.46}$$

This probability is, as expected, proportional to the number n of present photons.

In the *emission* process (Figs. 9.2 middle and bottom), using a similar argument, since the number of photons is changed from n to $n + 1$, one considers the action on the state $|n\rangle$ of the photon creation operator :

$$a^+|n\rangle = \sqrt{n+1}\,|n+1\rangle \qquad (9.47)$$

The emission probability per unit time P_e is proportional to the number of atoms N_1 that can emit a photon, to the square of the matrix element of this transition and is given by

$$P_e = N_1\,|\langle 0|V|1\rangle|^2\,|\langle n+1|a^+|n\rangle|^2 \qquad (9.48)$$

$$= N_1\,|\langle 0|V|1\rangle|^2\,(n+1) \qquad (9.49)$$

We find here that P_e is not zero even when $n = 0$ (*spontaneous* emission), but that the presence of photons increases P_e (*induced or stimulated* emission) : another way to express this property is to say that the presence of photons "stimulates" the emission of other photons. The induced emission was introduced by A. Einstein in 1916 and is at the origin of the *laser effect*. Remember that the name "laser" stands for Light Amplification by Stimulated Emission of Radiation.

In equilibrium at the temperature T, $\overline{n}$ photons of energy $h\nu$ are present : the absorption and emission processes exactly compensate, thus

$$N_0\overline{n} = N_1(\overline{n} + 1) \qquad (9.50)$$

Besides, for the atoms in equilibrium at T, the Maxwell-Boltzmann statistics allows one to write

$$\frac{N_1}{N_0} = e^{-\beta h\nu} \qquad (9.51)$$

One then deduces the value of $\overline{n}$

$$\overline{n} = \frac{1}{e^{\beta h\nu} - 1} \qquad (9.52)$$

which is indeed the Bose-Einstein distribution of photons. This yields $u(\nu, T)$ through the same calculation of the photons modes as above [formulae (9.29) to (9.33)].

This approach, which is based on advanced Quantum Mechanical results, has the advantage of introducing the stimulated emission concept and the description of an equilibrium from a balance between two phenomena.

9.2.5   Photometric Measurements : Definitions

We saw that a "black body" is emitting radiation corresponding to the thermal equilibrium in the cavity at the temperature T. A few simple geometrical considerations allow us to relate the total power emitted by a black body to its temperature and to introduce the luminance, which expresses the "color" of a heated body : we will then be able to justify that all black bodies at the same given temperature T are characterized by the same physical parameter and thus that one can speak of THE black body.

Let us first calculate the *total power P emitted* through the black-body aperture in the half-space outside the cavity. We consider the photons of frequency ν, to $d\nu$, which travel through the aperture of surface dS and have a velocity directed toward the angle θ to $d\omega$ with the normal $\vec{n}$ to dS. The number of these photons which pass the hole during dt are included in a cylinder of basis dS and height $c\cos\theta dt$ (Fig. 9.3). They carry an energy

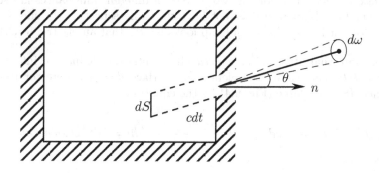

FIG. 9.3: Radiation emitted through an aperture of surface dS in a cone $d\omega$ around the direction θ.

$$d^2 P \cdot dt = \frac{d\omega}{4\pi} c \cos\theta \, dt \, dS \, u(\nu) d\nu \tag{9.53}$$

The total power P radiated into the half-space, by unit time, through a hole of unit surface is given by

$$P = \int_0^{\frac{\pi}{2}} \cos\theta \sin\theta d\theta \int_0^{2\pi} \frac{d\varphi}{4\pi} \int_0^\infty c \, u(\nu, T) d\nu \tag{9.54}$$

$$P = \frac{c}{4} \frac{U}{\Omega} = \frac{\pi^2}{60} \frac{(k_B T)^4}{\hbar^3 c^2} = \sigma T^4 \tag{9.55}$$

This power varies like T^4, this is the Stefan-Boltzmann law. The Stefan constant σ is equal to 5.67×10^{-8} watts $\cdot$ m$^{-2}\cdot$ kelvins^{-4}. At 6000 K, the approximate temperature of the sun surface, one gets $P = 7.3 \times 10^7$ W$\cdot$ m^{-2} ; at 300 K one finds 456 W$\cdot$m^{-2}.

Another way to handle the question is to analyze the *collection process of the radiation* issued from an arbitrary source by a detector. It is easy to

understand that the collected signal first depends on the spectral response of
the detector : for instance, a very specific detector is our eye, which is only
sensitive between 0.4 μm and 0.70 μm, with a maximum of sensitivity in the
yellow-green at 0.55 μm. Therefore specific units have been defined in *visual*
photometry, which take into account the physiological spectral response.

But the signal also depends on geometrical parameters, such as the detector
surface dS', the solid angle under which the detector sees the source or, equi-
valently, the source surface dS and the source-detector distance together with
their respective directions (Fig. 9.4).

A physical parameter, the *luminance L*, characterizes the source : in the visible
detection range the colors of two sources of different luminances are seen as
different (a red-hot iron does not have the same aspect as the sun or as an
incandescent lamp ; a halogen lamp looks whiter that an electric torch).

To be more quantitative, in *energetic* photometry one defines the infinitesimal
power d^2P received by the elementary surface dS' of a detector from the
surface dS of a source (Fig. 9.4) by the equation

$$d^2 P = LdS \cos\theta d\omega d\nu = LdS\frac{\cos\theta\cos\theta'}{r^2}dS'd\nu = LdS'\cos\theta' d\omega' d\nu \quad (9.56)$$

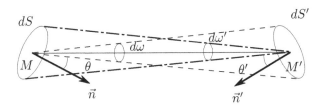

FIG. 9.4: Definition of the geometrical parameters of the source and detector.

The surface element of the source is centered on M, the detection element on
M', with $MM' = r$; θ is the angle between MM' and the normal $\vec{n}$ to dS,
$d\omega$ the solid angle under which the detector is seen from the source dS. One
notices that the definition is fully symmetrical and that alternatively one can
introduce dS', the solid angle $d\omega'$ under which the source is seen from the
detector, the angle θ' between $M'M$ and the normal to dS' in M'.

The luminance depends of T, ν and possibly of θ, the direction of the emission,
that is, $L(\nu, T, \theta)$.

When the source is a *black body* of surface dS in thermal equilibrium at the
temperature T, the power emitted toward dS' in the frequency range $(\nu, \nu+d\nu)$

is calculated from (9.53)

$$d^2 P = \frac{c}{4\pi} u(\nu, T) \, dS \cos\theta \, d\omega \, d\nu \qquad (9.57)$$

The comparison between (9.56) and (9.57) allows one to express the black body luminance L_0 versus $u(\nu, T)$:

$$L_0(\nu, T) = \frac{c}{4\pi} u(\nu, T) \qquad (9.58)$$

Thus *all black bodies at the same temperature T have the same luminance*, which allows one to define *the* black body. It is both the ideal source and detector since, once the equilibrium has been reached, it absorbs any radiation that it receives and re-emits it. *Its luminance does not depend on the direction θ of emission.* Formulae (9.56) and (9.57) also express that the surface which comes into play is the apparent surface, i.e., the projection $dS \cos\theta$ of the emitting surface on the plane perpendicular to MM'. For this whole surface the luminance is uniform and equal to $\frac{c}{4\pi} u(\nu, T)$: this is the Lambert law which explains why the sun, a sphere radiating like the black body, appears to us like a planar disk.

The heated bodies around us are generally not strictly black bodies, in the meaning that (9.58) is not fulfilled : their luminance$(L(\nu, T, \theta))$ not only depends on the frequency and the temperature, but also on the emission direction. Besides, we saw that the black body is "trapping" any radiation which enters the cavity : it thus behaves like a perfect absorber, which is not the most general case for an arbitrary body.

9.2.6 Radiative Balances

Consider an arbitrary body, exchanging energy only through radiation and having reached *thermal equilibrium* : one can only express a balance stating that the *absorbed power* is *equal to the emitted power* (in particular a transparent body neither absorbs nor emits energy). Assume that the studied body is illuminated by a black body and has an absorption coefficient $a(\nu, T, \theta)$ (equal to unity for the black body). If the studied body does not diffuse light, that is, radiation is only reemitted in the image direction of the source, and if it remains at the frequency it was absorbed, one obtains the relation

$$\underset{\text{absorption}}{L_0(\nu, T)a(\nu, T, \theta)} = \frac{c}{4\pi} u(\nu, T)a(\nu, T, \theta) = \underset{\text{emission}}{L(\nu, T, -\theta)} \qquad (9.59)$$

after simplifying the geometrical factors.

Relation (9.59) links the black body luminance $L_0(\nu, T)$ to that of the considered body, of absorption coefficient $a(\nu, T, \theta)$. The black body luminance is larger than, or equal to, that of any thermal emitter at the same temperature, *the black body is the best thermal emitter*. This is the Kirchhoff law.

In the case of a diffusing body (reemitting radiation in a direction other than $-\theta$) or of a fluorescent body (reemitting at a frequency different from that of its excitation), one can no longer express a detailed balance, angle by angle or frequency by frequency, between the ambient radiation and the body radiation. However, in steady-state regime, the total absorbed power is equal to the emitted power and the body temperature remains constant. The applicable relation is deduced from (9.59) after integration over the angles and/or over the frequencies.

One can also say that the power, assumed to be emitted by a black body, received by the studied body at thermal equilibrium at T, is either absorbed (and thus reemitted since T remains constant), or reflected. For instance for a diffusing body one writes :

$$received\ power = emitted\ power + reflected\ power$$

which implies for the luminances :

$$L_0(\nu, T) = L(\nu, T) + [L_0(\nu, T) - L(\nu, T)]$$
$$L_0(\nu, T) = L(\nu, T) + [1 - a(\nu, T)]L_0(\nu, T) \tag{9.60}$$

This is another way to obtain (9.59).

These types of balances are comparisons between the body luminance L and the black body luminance L_0 : if the interpretation of the L_0 expression requires Quantum Mechanics, the comparisons of luminances are only based on balances and were discovered during the 19th century.

More generally speaking, the exchanges of energy are taking place according to three processes : *conduction, convection* and *radiation*. Whereas the first two mechanisms require a material support, radiation which occurs through photons can propagate through vacuum : it is thanks to radiation that we are in contact with the universe. Through the intersideral vacuum, stars are illuminating us and we can study them by analyzing the received radiation. Our star, the sun, is sending a power of the order of a kW to each 1 m^2 surface of earth. The researches on solar energy try to make the best use of this radiation, using either thermal sensors, or photovoltaic cells ("solar cells") (see §8.5.2) which directly convert it into electricity.

9.2.7   Greenhouse Effect

Before interpreting the radiative balance of the earth, let us consider the greenhouse effect which allows one to maintain inside a building, made of glass formerly, often of plastics nowadays, a temperature higher than the one outside and thus favors particular vegetables or flowers cultivation (Fig. 9.5). The origin of the effect arises from the difference in absorption properties of the building walls for the visible and infrared radiations.

In the visible range, where the largest part of the radiation emitted by the sun is found, the walls are transparent, so that the sun radiation can reach the ground, where it is absorbed. The ground in turn radiates like a black body, with its frequency maximum in the infrared [Wien law, equation (9.41)].

Now the walls are absorbing in the infrared, they reemit a part of this radiation toward the ground, the temperature of which rises, and so on. At equilibrium the ground temperature is thus T_G. If a steady state has been reached, the power getting into the greenhouse, i.e., P_0, is equal to the one that is getting out of it :

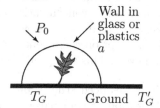

FIG. 9.5: Principle of the greenhouse effect.

$$P_0 = \sigma S T_G^4 (1 - a)$$

if S is the ground surface and a the absorption coefficient, by the walls, of the ground infrared radiation. In the absence of greenhouse, the ground temperature would be T'_G, such that

$$P_0 = \sigma S T_G'^4$$

The temperature T_G inside the greenhouse is higher than T'_G since

$$T_G = \frac{T'_G}{(1 - a)^{1/4}}$$

Here we assumed a perfect transparency of the walls in the visible and no reflection by the walls both in the visible and the infrared.

"The greenhouse effect" related to a possible climate global warming is an example of radiative balance applied to the earth. The earth is surrounded by a layer of gases, some of which corresponding to molecules with a permanent dipolar moment, due to their unsymmetrical arrangement of atoms : this is the case for H_2O, CO_2, O_3, etc. Thus the vibration and rotation spectra of these molecules evidence an infrared absorption.

The sun illuminates the earth with thermal radiation, the maximum of its spectrum being located in the visible [Eq. (9.41)]; this radiation is warming the ground. The earth, of temperature around 300 K, reemits thermal radiation toward space, the frequency maximum of this reemission is in the infrared. A fraction of the earth radiation is absorbed by the gases around the earth, which in turn reemit a fraction of it, and so on. Finally, according to the same phenomenon as the one described above for a greenhouse, the earth receives a power per unit surface larger than the one it would have reached it if there had only been vacuum around it ; its average temperature is 288 K (instead of 255 K if it had been surrounded by vacuum).

The global climate change, possibly induced by human activity, is related to an increase in concentration of some greenhouse effects gases (absorbing fluorocarboned compounds, which are now forbidden for use, and mainly CO_2). If the absorption coefficient increases, the ground temperature will vary in the same way. This will produce effects on the oceans level, etc... [1]

More advanced studies on the greenhouse effect account for the spectra of absorption or transparency of the gases layer around the earth according to their chemical species, etc.

In our daily life, we are surrounded by radiating objects like heating radiators, incandescent lamps. On the contrary we wish other objects to maintain their energy without radiating : this is the case, in particular, for thermos bottles, Dewar cans. We foresee here a very vast domain with technological, ecological and political involvements !

---

[1] The Intergovernmental Panel on Climate Change, established by the World Meteorological Organization and the United Nations Environment program, is assessing the problem of potential global climate change. An introduction brochure can be downloaded from its website (www.ipcc.ch).

Summary of Chapter 9

The chemical potential μ of *material* non-interacting particles following the Bose-Einstein statistics is located below their fundamental state ε_1.

The occupation factor by material bosons of an energy level ε is equal to

$$f(\varepsilon) = \frac{1}{e^{\beta(\varepsilon-\mu)} - 1}$$

At given density N/Ω, below a temperature T_B, which is a function of N/Ω, the Bose condensation takes place : a macroscopic number of particles then occupies the fundamental state ε_1, that is, the same quantum state. The thermodynamical properties in this low temperature range are mainly due to the particles still in the excited states.

The superfluidity of helium 4, the superconducting state of some solids, are examples of systems in the condensed Bose state, which is observed for $T < T_B$.

Photons are bosons in non-conserved number, their Bose-Einstein distribution is given by

$$f(\varepsilon) = \frac{1}{e^{\beta\varepsilon} - 1}$$

The Planck law expresses the spectral density in energy or frequency of photons in an empty cavity of volume Ω, in the frequencies range $[\nu, \nu + d\nu]$:

$$dU = \Omega u(\nu)d\nu = \Omega \frac{8\pi h}{c^3} \frac{\nu^3 d\nu}{e^{\beta h\nu} - 1}$$

This law gives the spectral distribution of the "black body" radiation, i.e., of thermal radiation at temperature $T = 1/k_B\beta$.

The black body is the perfect thermal emitter, its luminance is proportional to $u(\nu)$. The thermal emission of a body heated at temperature T corresponds to a luminance at maximum equal to that of the black body.

The greenhouse effect arises from the properties of the walls enclosing the greenhouse : higher transparency in the visible and larger infrared absorption. This phenomenon is also applicable to the radiation balance of the earth.

General Method for Solving Exercises and Problems

At the risk of stating obvious facts, we recall here that the text of a statistical physics exercise or problem is very precise, contains all the required information and that **each word really matters.**

Therefore it is recommended to first read the text with the highest attention.

The logic of the present course will always be apparent, i.e, you have to ask yourself the two following questions, in the order given here :

1. What are the microstates of the studied system ?
2. Which among these microstates are achieved in the problem conditions ?

1. The first question is generally solved through **Quantum Mechanics**, which reduces to **Classical Mechanics** in special cases (the ideal gas, for example).

Mostly this is equivalent to the question : "What are the energy states of the system ?" In fact they are often specified in the text.

At this stage there are three cases :

- **distinguishable** particles (for example of fixed coordinates) : see the first part of the course
- **classical mobile** particles : see the chapter on the ideal gas
- **indistinguishable** particles : see the second part of the course, in which the study is limited to the case of independent particles.

2. The second question specifically refers to **Statistical Physics**.

According to the physical conditions of the problem one will work in the **adapted statistical ensemble** :

- isolated system, fixed energy and fixed particles number : *microcanonical ensemble*

- temperature given by an energy reservoir and fixed number of particles : *canonical ensemble*

- system in contact with an energy reservoir (which dictates its temperature) and a particles reservoir (which dictates its chemical potential) : *grand canonical ensemble*

The systems of **indistinguishable particles** obeying the Pauli principle and studied in the framework of this course are solved in the **grand-canonical ensemble**. Their chemical potential is then set in such a way that the **average** number of particles coincides with the **real** number of particles given by the physics of the system.

Units and physical constants

(From R. Balian, "From microphysics to macrophysics : methods and appli-cations of statistical physics", Springer Verlag Berlin (1991))

We use the international system (SI), legal in most countries and adopted by most international institutions. Its fundamental units are the meter (m), the kilogram (kg), the second (sec), the ampere (A), the kelvin (K), the mole (mol), and the candela (cd), to which we add the radian (rad) and the stera-dian (sr).

The derived units called by specific names are : the hertz ($Hz = s^{-1}$), the newton ($N = m\ kg\ sec^{-2}$), the pascal ($Pa = N\ m^{-2}$), the joule ($J = N\ m$), the watt ($W = J\ sec^{-1}$), the coulomb ($C = A\ s$), the volt ($V = W\ A^{-1}$), the farad ($F = C\ V^{-1}$), the ohm ($\Omega = V\ A^{-1}$), the siemens ($S = A\ V^{-1}$), the weber ($Wb = V\ s$), the tesla ($T = Wb\ m^{-2}$), the henry ($H = Wb\ A^{-1}$), the Celsius degree (C), the lumen ($lm = cd\ sr$), the lux ($lx = lm\ m^{-2}$), the becquerel ($Bq = sec^{-1}$), the gray ($Gy = J\ kg^{-1}$) and the sievert ($Sv = J\ kg^{-1}$).

The multiples and submultiples are indicated by the prefixes deca ($da = 10$), hecto ($h = 10^2$), kilo ($k = 10^3$); mega ($M = 10^6$), giga ($G = 10^9$), tera ($T = 10^{12}$), peta ($P = 10^{15}$), exa ($E = 10^{18}$); deci ($d = 10^{-1}$), centi ($c = 10^{-2}$), milli ($m = 10^{-3}$); micro ($\mu = 10^{-6}$), nano ($n = 10^{-9}$), pico ($p = 10^{-12}$), femto ($f = 10^{-15}$), atto ($a = 10^{-18}$).

Constants of electromagnetic units	$\mu_0 = 4\pi \times 10^{-7} \, \mathrm{N \, A^{-2}}$ (defines the ampere)
	$\varepsilon_0 = 1/\mu_0 c^2$, $1/4\pi\varepsilon_0 \simeq 9 \times 10^9 \, \mathrm{N \, m^2 \, C^{-2}}$
light velocity	$c = 299\,792\,458 \, \mathrm{m \, sec^{-1}}$ (defines the meter)
	$c \simeq 3 \times 10^8 \, \mathrm{m \, sec^{-1}}$
Planck constant	$h = 6.626 \times 10^{-34} \, \mathrm{J \, sec}$
Dirac constant	$\hbar = h/2\pi \simeq 1.055 \times 10^{-34} \, \mathrm{J \, sec}$
Avogadro number	$\mathcal{N} \simeq 6.022 \times 10^{23} \, \mathrm{mol^{-1}}$ (by definition,
	the mass of one mole of $^{12}\mathrm{C}$ is 12 g)
Atomic mass unit	$1 \, \mathrm{u} = 1 \, \mathrm{g}/\mathcal{N} \simeq 1.66 \times 10^{-27} \, \mathrm{kg}$
	(or dalton or amu)
neutron and proton masses	$m_\mathrm{n} \simeq 1.0014 m_\mathrm{p} \simeq 1.0088 \, \mathrm{u}$
electron mass	$m \simeq 1 \, \mathrm{u}/1823 \simeq 9.11 \times 10^{-31} \, \mathrm{kg}$
Elementary charge	$e \simeq 1.602 \times 10^{-19} \, \mathrm{C}$
Faraday constant	$\mathcal{N}e \simeq 96\,485 \, \mathrm{C \, mol^{-1}}$
Bohr magneton	$\mu_\mathrm{B} = e\hbar/2m \simeq 9.27 \times 10^{-24} \, \mathrm{J \, T^{-1}}$
nuclear magneton	$e\hbar/2m_\mathrm{p} \simeq 5 \times 10^{-27} \, \mathrm{J \, T^{-1}}$
Fine structure constant	$\alpha = \dfrac{e^2}{4\pi\varepsilon_0 \hbar c} \simeq \dfrac{1}{137}$
Hydrogen atom :	
Bohr radius	$a_0 = \dfrac{\hbar}{mc\alpha} = \dfrac{4\pi\varepsilon_0 \hbar^2}{me^2} \simeq 0.53 \, \text{Å}$
binding energy	$E_0 = \dfrac{\hbar^2}{2ma_0^2} = \dfrac{m}{2\hbar^2}\left(\dfrac{e^2}{4\pi\varepsilon_0}\right)^2$
	$\simeq 13.6 \, \mathrm{eV}$
Rydberg constant	$R_\infty = E_0/hc \simeq 109\,737 \, \mathrm{cm^{-1}}$
Boltzmann constant	$k_B \simeq 1.381 \times 10^{-23} \, \mathrm{J \, K^{-1}}$
molar constant of the gas	$\mathcal{R} = \mathcal{N}k_B \simeq 8.316 \, \mathrm{J \, K^{-1}mol^{-1}}$
Normal conditions :	
pressure	$1 \, \mathrm{atm} = 760 \, \mathrm{torr} = 1.01325 \times 10^5 \, \mathrm{Pa}$
temperature	Triple point of water : 273.16 K
	(definition of the kelvin) or 0.01C
	(definition of the Celsius scale)
molar volume	$22.4 \times 10^{-3} \mathrm{m^3} \, \mathrm{mol^{-1}}$
Gravitation constant	$G \simeq 6.67 \times 10^{-11} \, \mathrm{m^3 \, kg^{-1} \, sec^{-2}}$
gravity acceleration	$g \simeq 9.81 \, \mathrm{m \, sec^{-2}}$
Stefan constant	$\sigma = \dfrac{\pi^2 k^4}{60\hbar^3 c^2}$
	$\simeq 5.67 \times 10^{-8} \, \mathrm{W \, m^{-2} \, K^{-4}}$

Definition of the photometric units	A light power of 1 W, at the frequency of 540 THz, is equivalent to 683 lm

Energy units and equivalences :	1 erg $= 10^{-7}$ J (not IS) 1 kWh$=3.6\times10^6$ J
electrical potential (electron-volt)	1 eV $\leftrightarrow 1.602 \times 10^{-19}$ J $\leftrightarrow$ 11 600 K
heat (calorie)	1 cal $= 4.184$ J (not SI; specific heat of 1 g water)
chemical binding	23 kcal mol^{-1} $\leftrightarrow$ 1 eV (not SI)
temperature $(k_B T)$	290 K $\leftrightarrow \frac{1}{40}$ eV (standard temperature)
mass (mc^2)	9.11×10^{-31} kg $\leftrightarrow$ 0.511 MeV (electron at rest)
wave number (hc/λ)	109 700 cm^{-1} $\leftrightarrow$ 13.6 eV (Rydberg)
frequency $(h\nu)$	3.3×10^{15} Hz $\leftrightarrow$ 13.6 eV

It is useful to remember these equivalencies to quickly estimate orders of magnitude.

Various non-SI units	1 ångström (Å)$=10^{-10}$ m (atomic scale) 1 fermi (fm)$=10^{-15}$ m (nuclear scale) 1 barn (b)$=10^{-28}$ m^2 1 bar$=10^5$ Pa 1 gauss (G)$=10^{-4}$ T 1 marine mile$=1852$ m 1 knot$=$1 marine mile per hour $\simeq 0.51$ m sec^{-1} 1 astronomical unit (AU) $\simeq 1.5 \times 10^{11}$ m (Earth-Sun distance) 1 parsec (pc) $\simeq 3.1 \times 10^{16}$ m (I UA/sec arc) 1 light-year $\simeq 0.95 \times 10^{16}$ m

Data on the Sun	Radius 7×10^8 m$=109$ earth radii Mass 2×10^{30} kg Average density 1.4 g cm^{-3} Luminosity 3.8×10^{26} W

A few useful formulae

Normalization of a Gaussian function :

$$\int_{-\infty}^{+\infty} dx \ e^{-ax^2} = \sqrt{\frac{\pi}{a}}$$

The derivation of this formula with respect to a yields the moments of the Gauss distribution.

Euler gamma function :

$$\Gamma(t) \equiv \int_0^\infty x^{t-1} e^{-x} dx = (t-1)\Gamma(t-1)$$

$$\int_0^\infty (1-x)^{s-1} x^{t-1} dx = \frac{\Gamma(s)\Gamma(t)}{\Gamma(s+t)}$$

$$\Gamma(t)\Gamma(1-t) = \frac{\pi}{\sin \pi t} \quad , \quad \Gamma\left(\frac{1}{2}\right) = \sqrt{\pi}$$

Stirling formula :

$$t! = \Gamma(t+1) \underset{t\to\infty}{\sim} t^t \ e^{-t} \sqrt{2\pi t}$$

Binomial series :

$$(1+x)^t = \sum_{n=0}^\infty \frac{x^n}{n!} \frac{\Gamma(t+1)}{\Gamma(t+1)-n)} = \sum_{n=0}^\infty \frac{(-x)^n}{n!} \frac{\Gamma(n-t)}{\Gamma(-t)} \quad , \quad |x| < 1$$

Poisson formula :

$$\sum_{n=-\infty}^{+\infty} f(n) = \sum_{l=-\infty}^{+\infty} \tilde{f}(2\pi l) \equiv \sum_{l=-\infty}^{+\infty} \int_{-\infty}^{+\infty} dx f(x) e^{2\pi i l x}$$

Euler-Maclaurin formula :

$$\frac{1}{\varepsilon} \int_a^{a+\varepsilon} dx f(x) \approx \frac{1}{2}[f(a) + f(a+\varepsilon)] - \frac{\varepsilon}{12} f'(x) \mid_a^{a+\varepsilon} + \frac{\varepsilon^3}{720} f'''(x) \mid_a^{a+\varepsilon} + \dots$$

$$\approx f(a + \tfrac{1}{2}\varepsilon) + \frac{\varepsilon}{24} f'(x) \mid_a^{a+\varepsilon} - \frac{7\varepsilon^3}{5760} f'''(x) \mid_a^{a+\varepsilon} + \dots$$

This formula allows one to calculate the difference between an integral and a sum over n, for $a = n\varepsilon$.

Constants

$$e \simeq 2.718, \qquad \pi \simeq 3.1416$$

$$\gamma \equiv \lim(1 + \dots + \frac{1}{n} - \ln n) \simeq 0.577 \quad \text{Euler constant.}$$

Riemann zeta function :

$$\zeta(t) \equiv \sum_{n=1}^{\infty} \frac{1}{n^t}, \qquad \int_0^{\infty} \frac{x^{t-1} dx}{e^x - 1} = \Gamma(t)\zeta(t)$$

$$\int_0^{\infty} \frac{x^{t-1} dx}{e^x + 1} = (1 - 2^{-t+1})\Gamma(t)\zeta(t)$$

t	1.5	2	2.5	3	3.5	4	5
ζ	2.612	$\frac{1}{6}\pi^2$	1.341	1.202	1.127	$\frac{1}{90}\pi^4$	1.037

Dirac distribution :

$$\frac{1}{2\pi} \int_{-\infty}^{+\infty} dx\, e^{ixy/a} = \delta(y/a) = |a|\delta(y)$$

$$\frac{1}{|a|} \sum_{\ell=-\infty}^{+\infty} e^{2\pi i \ell y/a} = \delta(y/a)$$

$$\lim_{t\to\infty} \frac{\sin tx}{x} = \lim_{t\to\infty} \frac{1 - \cos tx}{tx^2} = \pi\delta(x)$$

$$f(x)\delta(x) = f(0)\delta(x) \quad, \quad f(x)\delta'(x) = -f'(0)\delta(x) + f(0)\delta'(x)$$

If $f(x) = 0$ at the $x = x_i$ points, one has

$$\delta[f(x)] = \sum_i \frac{1}{|f'(x_i)|}\delta(x - x_i)$$

Exercises and Problems

A course can only be fully understood after practice through exercises and problems. There are many excellent collections of classical exercises in Statistical Physics. In what follows, we reproduce original texts which were recently given as examinations at Ecole Polytechnique. An examination of Statistical Physics at Ecole Polytechnique usually consists in an exercise and a problem. As examples, you will find here the exercises given in 2000, 2001 and 2002 and the problems of 2001 and 2002 which were to be solved by the students who had attended the course presented in this book.

Electrostatic Screening : exercise 2000
Magnetic Susceptibility of a "Quasi-1D" Conductor : exercise 2001
Entropies of the $HC\ell$ Molecule : exercise 2002
Quantum Boxes and Optoelectronics : problem 2001
Physical Foundations of Spintronics : problem 2002

Exercise 2000 : Electrostatic Screening

I. Let us consider N free fermions of mass m_e and spin $1/2$, without any mutual interaction, contained in a macroscopic volume Ω maintained at temperature T.

I.1. Recall the equation relating N/Ω to the chemical potential μ of these fermions through the density of states in $\vec{k}$ (do not calculate the integral).

I.2. In the high temperature and low density limit recall the **approximate** integral expression relating N/Ω to μ (now calculate the integral).

I.3. The same fermions, at high temperature and small density, are now submitted to a potential energy $V(\vec{r})$. Express the total energy of one such particle versus $\vec{k}$. Give the **approximate** expression relating N to μ and $V(\vec{r})$.

I.4. Deduce that the volume density of fermions is given by

$$n(\vec{r}) = n_0 \exp(-\beta V(\vec{r})) \quad ,$$

where $\beta = 1/k_B T$ and n_0 is a density that will be expressed as an integral.

I.5. Show that at very small potential energy, such that $\beta V(\vec{r}) \ll 1$ for any $\vec{r}$, then $n_0 = N/\Omega$.

II. From now on, we consider an illuminated semiconductor, of volume Ω, maintained at temperature T. It contains N electrons, of charge $-e$, in its conduction band together with $P = N$ holes (missing electrons) in its valence band. The holes will be taken as fermions, of spin $1/2$, mass m_h and charge $+e$. We assume that the high temperature limit is realized.

Assume that a fixed point charge Q is present at the coordinates origin, which, if alone, would create an **electrostatic potential** :

$$\Phi_{\text{ext}}(\vec{r}) = \frac{Q}{4\pi\varepsilon_0\varepsilon_r r}$$

where ε_r is the relative dielectric constant of the considered medium.

In fact, the external charge Q induces inhomogeneous charge densities $-en(\vec{r})$ and $+ep(\vec{r})$ inside both fermion gases. These induced charges in turn generate electrostatic potentials, which modify the densities of mobile charges, and so on.

You are going to solve this problem by a self-consistent method in the following way : let $\Phi_{\text{tot}}(\vec{r})$ be the **total electrostatic potential**, acting on the particles.

II.1. Express the potential energies, associated with $\Phi_{tot}(\vec{r})$, to which the electrons and the holes are respectively subjected. Deduce from I. the expressions $n(\vec{r})$ and $p(\vec{r})$ of the corresponding densities of mobile particles.

II.2. The electrostatic potential $\Phi_{tot}(\vec{r})$ is a solution of the Poisson equation, in which the charge density includes all the charges of the problem. Admit that the charge density $Q\delta(\vec{r})$ is associated with the point charge Q located at $\vec{r}$, where $\delta(\vec{r})$ is the Dirac delta distribution.

Show that, in the high temperature limit and to the lowest order in $\Phi_{tot}(\vec{r})$, the total electrostatic potential is solution of

$$\Delta\Phi_{tot}(\vec{r}) = \frac{\Phi_{tot}(\vec{r})}{\lambda^2} - \frac{Q\delta(\vec{r})}{\varepsilon_0\varepsilon_r}$$

where λ is a constant that will be expressed versus N/Ω and T. What is the unit of λ?

II.3. Estimate λ for $N/\Omega = 10^{22}$ m^{-3}, $T = 300$ K, $\varepsilon_r = 12.4$.

II.4. The solution of this equation is

$$\Phi_{tot}(\vec{r}) = \frac{Q\exp(-r/\lambda)}{4\pi\varepsilon_0\varepsilon_r r}$$

Indicate the physical implications of this result in a few sentences.

Exercise 2001 : Magnetic Susceptibility of a "Quasi-1D"

I. Consider an assembly of N magnetic ions at temperature T inside a volume Ω. These ions are independent and distinguishable. The electronic structure of an ion consists in a fundamental non-degenerate level $|0\rangle$, of energy ε_0, and a doubly degenerate excited state of energy $\varepsilon_0 + \lambda$. Under the effect of a static and uniform magnetic field $\mathbf{B}$, the fundamental level remains at the energy ε_0 while the excited level splits into two sublevels $|+\rangle$ and $|-\rangle$ of respective energies $\varepsilon_0 + \lambda + \gamma B$ and $\varepsilon_0 + \lambda - \gamma B$.

I.1. Without any calculation, plot the location of the states $|0\rangle$, $|+\rangle$ and $|-\rangle$ versus the applied magnetic field, in the following special cases :

   a) $\lambda = 0$, $\quad k_B T \gg \gamma B$
   b) $\lambda \gg k_B T \gg \gamma B$
   c) $k_B T \gg \lambda \gg \gamma B$
   d) $\gamma B \gg \lambda \gg k_B T$

In each case indicate which states are occupied at temperature T.

I.2. Calculate the system free energy, assuming that the only degrees of freedom are those, of electronic origin, related to the three levels $|0\rangle$, $|+\rangle$ and $|-\rangle$. Deduce the algebraic expressions of the system magnetization $M = -\dfrac{\partial F}{\partial B}$ and of its susceptibility $\chi = \lim_{B \to 0} \dfrac{M}{\Omega B}$ (Ω is the volume).

I.3. On the expressions of the magnetization and of the susceptibility obtained in I.2. *only* discuss the limit cases a), b), c), d) of question I.1 (take $\dfrac{\gamma B}{k_B T} = x$). Verify that the obtained results are in agreement with the qualitative considerations of I.1.

I.4. The figure below represents the variation versus temperature of the paramagnetic susceptibility of the organic "quasi-one-dimensional" compound $[\text{HN}(\text{C}_2\text{H}_5)_3](\text{TCNQ})_2$. Using the results of the present exercise can you *qualitatively* interpret the shape of this curve?

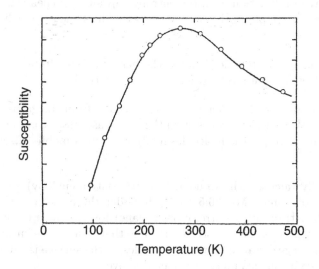

FIG. 1: Magnetic susceptibility of the quasi-one-dimensional organic compound versus temperature.

II. Let a paramagnetic ion have a total spin S. You will admit that, in presence of a static and uniform magnetic field $\vec{B}$, the levels of this ion take the energies $\varepsilon_l = l\gamma B$, where $-S \le l \le S$, with $l = -S, (-S+1), (-S+2), \ldots + S$.

Consider an ion at thermal equilibrium at the given temperature T. Calculate its free energy f and its average magnetic moment $m = -\dfrac{\partial f}{\partial B}$.

Write the expressions of the magnetization in the following limit cases : high and low magnetic fields; high and low temperatures. In each situation give a brief physical comment.

Exercise 2002 : Entropies of the HCℓ Molecule

1. Let a *three-dimensional* ideal gas be made of N monoatomic molecules of mass m, confined inside the volume Ω and at temperature T. **Without demonstration** recall the expressions of its entropy S_{3D} and of its chemical potential μ_{3D}.

2. For a *two-dimensional* ideal gas of N' molecules, mobile on a surface A, of the same chemical species as in 1., give the expressions, versus A, m and T, of the partition function Z_{2D}, the entropy S_{2D} and the chemical potential μ_{2D}. To obtain these physical parameters, follow the same approach as in the course.

Which of these parameters determines whether a molecule from the ideal gas of question 1. will spontaneously adsorb onto surface A?

3. Let us now assume that each of the N molecules from the gas is adsorbed on surface A. What is, with respect to the gas phase, the variation per mole of translation entropy if the molecules now constitute a mobile adsorbed film on A?

4. Numerically calculate the value of S_{3D} (translation entropy) for one mole of HCℓ, of molar mass M $= 36.5$, at 300K under the atmospheric pressure. Calculate S_{2D} (translation entropy on a plane) for one mole of such a gas, adsorbed on a porous silicon surface : assume that, due to the surface corrugation, the average distance between molecules on the surface is equal to $1/30$ of the distance in the gas phase considered above.

5. In fact, in addition to its translation degrees of freedom, the HCℓ molecule possesses rotation degrees of freedom. Write the general expression of the partition function of N molecules of this gas versus z_{tr} and z_{rot}, the translation and rotation partition functions for a single molecule, accounting for the fact that the molecules are indistinguishable. The expressions of z_{tr} and z_{rot} will be calculated in the next questions.

6. In this question one considers a molecule constrained to rotate in a plane around a fixed axis. Its inertia moment with respect to this axis is I_A. The rotation angle ϕ varies from 0 to 2π, the classical momentum associated with ϕ is p_ϕ. The hamiltonian for one molecule corresponding to this motion is given by

$$h = \frac{p_\phi^2}{2I_A}$$

The differential elementary volume of the corresponding phase space is $d\phi\, dp_\phi/h$, where h is the Planck constant. Calculate the partition function $z_{1\text{rot}}$.

7. In the case of the linear $HC\ell$ molecule, of inertia moment I, free to rotate in space, the rotation hamiltonian for a single molecule is written, in spherical coordinates,

$$h = \frac{p_\theta^2}{2I} + \frac{p_\phi^2}{2I\sin^2\theta}$$

Here θ is the polar angle $(0 \leq \theta \leq \pi)$, ϕ the azimuthal angle $(0 \leq \phi \leq 2\pi)$; p_θ (respectively p_ϕ) is the momentum associated with θ (respectively ϕ). The volume element of the phase space is now equal to $d\phi\, dp_\phi\, d\theta\, dp_\theta/h^2$. Calculate the corresponding partition function $z_{2\text{rot}}$.

Show that the dependence in temperature and in inertia moment obtained for $z_{2\text{rot}}$ is the expected one, when this expression of the rotation energy is taken into account. Verify that the expression found for $z_{2\text{rot}}$ is in agreement with the one given in the course for a linear molecule in the classical approximation (be careful, the $HC\ell$ molecule is not symmetrical).

8. Numerically calculate the rotation entropy by mole at 300K of a $HC\ell$ mole.

Take $I = 2.7 \times 10^{-40}$ g $\cdot$ cm ^{2}.

Problem 2001 : Quantum Boxes and Optoelectronics

For more clarity, comments are written in italics and questions in standard characters.

Semiconducting quantum boxes are much studied nowadays, because of their possible applications in optoelectronics. The aim of the present problem is to study the rate of spontaneous emission and the radiative efficiency in quantum boxes : these are two important physical parameters for the elaboration of emission devices (quantum–boxes lasers). We will particularly focus on the experimental dependence of these parameters.

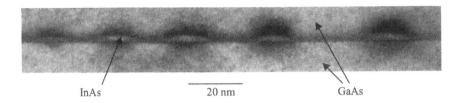

InAs   20 nm   GaAs

FIG. 1 : Electron microscope image of quantum boxes of InAs in GaAs.

Quantum boxes are made of three-dimensional semiconducting inclusions of nanometer size (see Fig.1). In the studied case, the inclusions consist of InAs and the surrounding material of GaAs. The InAs quantum boxes confine the electrons to a nanometer scale in three dimensions and possess a discrete set of electronic valence and conduction states; they are often called "artificial atoms." In the following we will adopt a simplified description of their electronic structure, as presented below in Fig. 2.

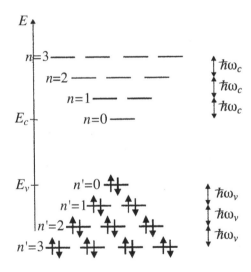

FIG. 2: Simplified model of the electronic structure of an InAs quantum box and of the filling of its states at thermal equilibrium at $T = 0$.

We consider that the energy levels in the conduction band, labeled by a positive integer n, are equally spaced and separated by $\hbar\omega_c$; the energy of the fundamental level $(n = 0)$ is noted E_c. The nth level is $2(n + 1)$ times degenerate, where the factor 2 accounts for the spin degeneracy and $(n + 1)$ for the orbital degeneracy. Thus a conduction state is identified through the set of three in-

teger quantum numbers $|n, l, s\rangle$, with $n > 0$ and $0 \leq l \leq n$, $s = \pm\dfrac{1}{2}$. The first valence states are described in a similar way, with analogous notations $(E_v, \hbar\omega_v, n', l', s')$.

I. Quantum Box in Thermal Equilibrium

At zero temperature and thermal equilibrium, the valence states are occupied by electrons, whereas the conduction states are empty. We are going to show in this section that the occupation factor of the conduction states of the quantum box in equilibrium remains small for temperatures up to 300K. Therefore, only in this part, the electronic structure of the quantum box will be roughly described by just considering the conduction and valence band levels lying nearest to the band gap. Fig.3 illustrates the assumed electronic configuration of the quantum box at zero temperature.

E_C ——

E_V

FIG. 3 : Electronic configuration at $T = 0$.

I.1. Give the average number of electrons in each of these two (spin-degenerate) levels versus $\beta = 1/k_B T$, where T is the temperature, and the chemical potential μ .

I.2. Find an implicit relation defining μ versus β, by expressing the conservation of the total number of electrons in the thermal excitation process.

I.3. Show that, in the framework of this model, μ is equal to $(E_c + E_v)/2$ independently of temperature.

I.4. Estimate the average number of electrons in the conduction level at 300K and conclude. For the numerical result one will take $E_c - E_v = 1$ eV.

In optoelectronic devices, like electroluminescent diodes or lasers, electrons are injected into the conduction band and holes into the valence band of the semiconducting active medium : the system is then out of equilibrium. The return toward equilibrium takes place through electron transitions from the conduction band toward the valence band, which may be associated with light emission.

Although the box is not in equilibrium, you will use results obtained in the course (for systems in equilibrium) to study the distribution of electrons in the conduction band, and that of holes in the valence band. **Consider that, inside a given band, the carriers are in equilibrium among themselves, with a specific chemical potential for each band.** *This is justified by the fact that the phenomena responsible for transitions between conduction states, or between valence states, are much faster than the recombination of electron-hole pairs.*

One now assumes in sections II and III that the quantum box exactly contains one electron-hole pair (regime of weak excitation) : the total number of electrons contained in all the conduction states is thus equal to 1 ; the same assumption is valid for the total number of holes present in the valence states.

II. Statistical Study of the Occupation Factor of the Conduction Band States

II.1. Write the Fermi distribution for a given electronic state $|n, l, s\rangle$ versus $\beta = 1/k_B T$ and μ, the chemical potential of the conduction electrons. Show that this distribution factor $f_{n,l,s}$ only depends of the index n; it will be written f_n in the following.

II.2. Expressing that the total number of electrons is 1, find a relation between μ and β.

II.3. Show that $\mu < E_c$.

II.4. Show that f_n is a decreasing function of n. Deduce that

$$f_n \leq \frac{1}{(n+1)(n+2)}$$

(*Hint* : give an upper bound and a lower bound for the population of electronic states of energy smaller or equal to that of level n).

II.5. Deduce that for states other than the fundamental one, it is very reasonable to approximate the Fermi-Dirac distribution function by a Maxwell-Boltzmann distribution and this independently of temperature. Deduce that the chemical potential is defined by the following implicit relation :

$$\frac{1}{2} = g(\beta, w) , \quad \text{with} \quad w = e^{\beta(\mu - E_c)}$$

$$\text{and} \quad g(\beta, w) = \frac{w}{w+1} + w \sum_{n \geq 1} (n+1) e^{-\beta n \hbar \omega_c} \quad (1)$$

II.6. Show that the function g defined in II.5 is an increasing function of w and a decreasing function of β. Deduce that the system fugacity w decreases when the temperature increases and that μ is a decreasing function of T.

II.7. Specify the chemical potential limits at high and low temperatures.

II.8. Deduce that beyond a critical temperature T_c – to be estimated only in the next questions – it becomes legitimate to make a Maxwell-Boltzmann-type approximation for the occupation factor of the fundamental level.

II.9. Now one tries to estimate T_c. Show that expression (1) can also be written under the form :

$$\frac{1}{2} = \frac{w}{w+1} - w + \frac{w}{(1-e^{-\beta\hbar\omega_c})^2} \tag{2}$$

[*Hint* : as a preliminary, calculate the sum $\chi(\beta) = \sum_{n\geq 1} e^{-\beta n\hbar\omega_c}$ and compare it to the sum entering into equation (1)].

II.10. Consider that the Maxwell-Boltzmann approximation is valid for the fundamental state if $f_0 \leq \dfrac{1}{e+1}$. Hence deduce T_c versus $\hbar\omega_c$.

II.11. For a temperature higher than T_c show that :

$$f_n = e^{-\beta n\hbar\omega_c}\frac{(1-e^{-\beta\hbar\omega_c})^2}{2} \tag{3}$$

III. Statistics of holes

III.1. Give the average number of **electrons** in a valence band state $|n', l', s'\rangle$ of energy $E_{n'}$, versus the chemical potential μ_v of the valence electrons.

(*Recall* : μ_v is different from μ because the quantum box is out of equilibrium when an electron-hole pair is injected into it.)

III.2. Deduce the average number of **holes** $h_{n',l',s'}$ present in this valence state of energy $E_{n'}$. Show that the holes follow a Fermi-Dirac statistics, if one associates the energy $-E_{n'}$ to this fictitious particle. Write the hole chemical potential versus μ_v.

It will thus be possible to easily adapt the results of section II, obtained for the conduction electrons, to the statistical study of the holes in the valence band.

IV. Radiative lifetime of the confined carriers

*Consider a quantum box containing **an electron-hole pair at the instant**
$t = 0$; several physical phenomena can produce its return to equilibrium. In the
present section IV we only consider the **radiative** recombination of the charge
carriers through spontaneous emission : the deexcitation of the electron from
the conduction band to the valence band generates the emission of a photon.*

*However an optical transition between a conduction state $|n, l, s\rangle$ and a valence
state $|n', l', s'\rangle$ is only allowed if specific selection rules are satisfied. For the
InAs quantum boxes, one can consider, to a very good approximation, that for
an allowed transition $n = n'$, $l = l'$ and $s = s'$ and that each allowed optical
transition has the same strength. Consequently, the probability per unit time
of radiative recombination for the electron-hole pair is simply given by*

$$\frac{1}{\tau_r} = \sum_{n,l,s} \frac{f_{n,l,s} \cdot h_{n,l,s}}{\tau_0} \tag{4}$$

*The time τ_r is called the radiative lifetime of the electron-hole pair ; the time
τ_0 is a constant characteristic of the strength of the allowed optical transitions.*

IV.1. Show that, if only radiative recombinations take place and if the box
contains one electron-hole pair at $t = 0$, the average number of electrons in
the conduction band can be written e^{-t/τ_r} for $t > 0$.

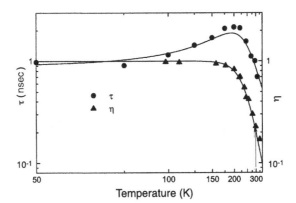

FIG. 4: Lifetime τ and radiative yield η measured for InAs quantum boxes ;
the curves are the result of the theoretical model developed in the present
problem. The scales of τ, η and T are logarithmic.

IV.2. Now the temperature is zero. Calculate the radiative lifetime using (4).
*The lifetime τ of the electron-hole pair can be measured. Its experimental
variation versus T is shown in Fig. 4.*

IV.3. Assume that at low temperature one can identify the lifetime τ and the radiative lifetime τ_r. Deduce τ_0 from the data of Fig. 4.

IV.4. From general arguments show that the radiative lifetime τ_r is longer for $T \neq 0$ than for $T = 0$. (Do not try to calculate τ_r in this question.)
Now the temperature is in the 100–300K range. For InAs quantum boxes, some studies show that the energy spacing $\hbar\omega$ between quantum levels is of the order of 15 meV for the valence band and of 100 meV for the conduction band.

IV.5. Estimate T_c for the conduction electrons and the valence holes using the result of II.10. Show that a zero temperature approximation is valid at room temperature for the distribution function of the conduction states. What is the approximation then valid for the valence states ?

IV.6. Deduce the dependence of the radiative lifetime versus temperature T for $T > T_c$.

IV.7. Compare to the experimental result for τ shown in Fig. 4. Which variation is explained by our model ? What part of the variation remains unexplained ?

IV.8. From the experimental values of τ at $T = 200$ K and at $T = 0$ K deduce an estimate of the energy spacing between hole levels. Compare to the value of 15 meV suggested in IV.4.

V. Non-Radiative Recombinations

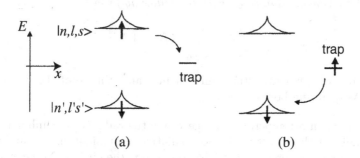

FIG. 5 : A simple two-step mechanism of non-radiative recombination.

*The electron-hole pair can also recombine without emission of a photon (**non-radiative recombination**). A simple two-step mechanism of non-radiative recombination is schematized as follows : (a) a trap located in the neighborhood of the quantum box captures the electron; (b) the electron then recombines with* the hole, without emission of a photon. The symmetrical mechanism relying on the initial capture of a hole is also possible. One expects that such a phenomenon will be the more likely as the captured electron (or hole) lies in an excited state of the quantum box, because of its more extended wave function of larger amplitude at the trap location.

The radiative and non-radiative recombinations are mutually independent. Write $1/\tau_{nr}$ for the probability per unit time of non-radiative recombination of the electron-hole pair; the total recombination probability per unit time is equal to $1/\tau$.

The dependence of the emission quantum yield η of the InAs quantum boxes is plotted versus T in the Fig. 4 above : η is defined as the fraction of recombinations occuring with emission of a photon.

V.1. Consider a quantum box containing an electron-hole pair at the instant $t = 0$. Write the differential equation describing the evolution versus time of the average number of electrons in the conduction band, assuming that the electron-hole pair can recombine either with or without emission of radiation.

V.2. Deduce the expressions of τ and η versus τ_r and τ_{nr}.

V.3. Comment on the experimental dependence of η versus T. Why is it consistent with the experimental dependence of τ ?

Assume that the carriers, electrons or holes, can be trapped inside the non-radiative recombination center only if they lie in a level of the quantum box of large enough energy (index n higher than or equal to n_0). Write the non-radiative recombination rate under the following form :

$$\frac{1}{\tau_{nr}} = \gamma \sum_{n \geq n_0, l, s} (f_{n,l,s} + t_{n,l,s}) \tag{5}$$

V.4. From the experimental results deduce an estimate of the index n_0 for the InAs quantum boxes.

[Hint : work in a temperature range where the radiative recombination rate is negligible with respect to the non-radiative recombination rate and write a simple expression versus $1/T$, to first order in η, then in $\ln \eta$. You can assume, and justify the validity of this approximation, that the degeneracy of any level of index higher than or equal to n_0 can be replaced by $2(n_0 + 1)$].

This simple description of the radiative and non-radiative recombination processes provides a very satisfactory understanding of the experimental dependences of the electron-hole pairs lifetime and of the radiative yield of the InAs *quantum boxes (see Fig. 4). These experimental results and their modeling are taken from a study performed at the laboratory of Photonics and Nano-structures of the CNRS (French National Center of Scientific Research) at Bagneux, France.*

Problem 2002 : Physical Foundations of Spintronics

A promising research field in material physics consists in using the electron spin observable as a medium of information (spintronics). This observable has the advantage of not being directly affected by the electrostatic fluctuations, contrarly to the space observables (position, momentum) of the electron. Thus, whereas in a standard semiconductor the momentum relaxation time of an electron does not exceed 10^{-12} sec, its spin relaxation time can reach 10^{-8} sec. Here we will limit ourselves to the two-dimensional motion of electrons (quantum well structure). The electron effective mass is m. The accessible domain is the rectangle $[0, L_x] \times [0, L_y]$ of the xOy plane. The periodic limit conditions will be chosen at the edge of this rectangle. *Part II can be treated if the results indicated in part I are assumed.*

Part I : Quantum Mechanics

Consider the motion of an electron described by the hamiltonian

$$\hat{H}_0 = \frac{\hat{\vec{p}}^2}{2m} + \alpha(\hat{\sigma}_x \hat{p}_x + \hat{\sigma}_y \hat{p}_y) \qquad \alpha \geq 0$$

where $\hat{\vec{p}} = -i\hbar\vec{\nabla}$ is the momentum operator of the electron and the $\hat{\sigma}_i$ ($i = x, y, z$) represent the Pauli matrices. Do not try to justify this hamiltonian. Assume that at any time t the electron state is factorized into a space part and a spin part, the space part being a plane wave of wave vector $\vec{k}$. This normalized state is thus written

$$\frac{e^{i\vec{k}\cdot\vec{r}}}{\sqrt{L_x L_y}}[a_+(t)|+\rangle + a_-(t)|-\rangle] = \frac{e^{i\vec{k}\cdot\vec{r}}}{\sqrt{L_x L_y}}\begin{pmatrix} a_+(t) \\ a_-(t) \end{pmatrix} \qquad (1)$$

$$\text{with } |a_+(t)|^2 + |a_-(t)|^2 = 1 \qquad (2)$$

where $|\pm\rangle$ are the eigenvectors of $\hat{\sigma}_z$. One will write $k_x = k\cos\phi$, $k_y = k\sin\phi$, with $k \geq 0$, $0 \leq \phi < 2\pi$.

I.1. Briefly justify that one can indeed search a solution of the Schroedinger equation under the form given in (2), where k and ϕ are time-independent. Write the evolution equations of the coefficients $a_\pm(t)$.

I.2

(a) For fixed k and ϕ, show that the evolution of vector $\begin{pmatrix} a_+(t) \\ a_-(t) \end{pmatrix}$ des-

cribing the spin state can be put under the form : $i\hbar\frac{d}{dt}\begin{pmatrix} a_+(t) \\ a_-(t) \end{pmatrix} =$

$\hat{H}_s \begin{pmatrix} a_+(t) \\ a_-(t) \end{pmatrix}$ where $\hat{H}_s$ is a 2×2 hermitian matrix, that will be explicitly written versus k, ϕ, and the constants α, $\hbar$ and m.

(b) Show that the two eigenenergies of the spin hamiltonian $\hat{H}_s$ are : $E_\pm = \dfrac{\hbar^2 k^2}{2m} \pm \alpha\hbar k$ with the corresponding eigenstates :

$$|\chi_+\rangle = \frac{1}{\sqrt{2}}\begin{pmatrix} 1 \\ e^{i\phi} \end{pmatrix} \qquad\qquad |\chi_-\rangle = \frac{1}{\sqrt{2}}\begin{pmatrix} 1 \\ -e^{i\phi} \end{pmatrix} \qquad (3)$$

(c) Deduce that the set of $|\Psi_{\vec{k},\pm}\rangle$ states :

$$|\Psi_{\vec{k},+}\rangle = \frac{e^{i\vec{k}\cdot\vec{r}}}{\sqrt{2L_xL_y}}\begin{pmatrix} 1 \\ e^{i\phi} \end{pmatrix} \qquad |\Psi_{\vec{k},-}\rangle = \frac{e^{i\vec{k}\cdot\vec{r}}}{\sqrt{2L_xL_y}}\begin{pmatrix} 1 \\ -e^{i\phi} \end{pmatrix} \qquad (4)$$

with $k_i = 2\pi n_i/L_i$ ($i = x, y$ and n_i positive, negative or null integers) constitute a basis, well adapted to the problem, of the space of the one-electron states.

I.3 Assume that at $t = 0$ the state is of the type (2) and write $\omega = \alpha k$.

(a) Decompose this initial state on the $|\Psi_{\vec{k},\pm}\rangle$ basis.

(b) Deduce the expression of $a_\pm(t)$ versus $a_\pm(0)$.

I.4 One defines $s_z(t)$, the average value at time t of the z-component of the electron spin $\hat{S}_z = (\hbar/2)\,\hat{\sigma}_z$. Show that :

$$s_z(t) = s_z(0)\,\cos(2\omega t) + \hbar\,\sin(2\omega t)\,\mathrm{Im}\left(a_+(0)^*\,a_-(0)\,e^{-i\phi}\right)$$

I.5 The system is installed in a magnetic field $\vec{B}$ lying along the z-axis. The effect of $\vec{B}$ on the orbital variables is neglected. The hamiltonian thus becomes $\hat{H} = \hat{H}_0 - \gamma B\hat{S}_z$.

(a) Why does the momentum $\vec{k}$ remain a good quantum number ? Deduce that, for a fixed $\vec{k}$, a spin hamiltonian $\hat{H}_s^{(B)}$ can still be defined. You will express it versus $\hat{H}_s$, γ, B and $\hat{\sigma}_z$.

(b) Show that the eigenenergies of $\hat{H}_s^{(B)}$ are

$$E_\pm^{(B)} = \frac{\hbar^2 k^2}{2m} \pm \hbar\sqrt{\alpha^2 k^2 + \gamma^2 B^2/4}$$

(c) Consider a weak field $\vec{B}$ $(\gamma B \ll \alpha k)$. The expression of the eigenvectors of $\hat{H}_s$ to first order in B are then given by

$$|\chi_+^{(B)}\rangle = |\chi_+\rangle - \frac{\gamma B}{4\alpha k}|\chi_-\rangle \qquad |\chi_-^{(B)}\rangle = |\chi_-\rangle + \frac{\gamma B}{4\alpha k}|\chi_+\rangle$$

Deduce that the average values of $\hat{S}_z$ in the states $|\chi_\pm^{(B)}\rangle$ are

$$\langle\chi_+^{(B)}|\hat{S}_z|\chi_+^{(B)}\rangle = -\frac{\hbar\gamma B}{4\alpha k} \qquad \langle\chi_-^{(B)}|\hat{S}_z|\chi_-^{(B)}\rangle = +\frac{\hbar\gamma B}{4\alpha k}$$

(d) What are the values of $|\chi_\pm^{(B)}\rangle$ and of the matrix elements $\langle\chi_\pm^{(B)}|\hat{S}_z|\chi_\pm^{(B)}\rangle$, in the case of a strong field $\vec{B}$ (such that $\gamma B \gg \alpha k$).

Part II : Statistical Physics

In this part, we are dealing with the properties of an assembly of N electrons, assumed to be without any mutual interaction, each of them being submitted to the hamiltonian studied in Part I. We will first assume that there is no magnetic field; the energy levels are then distributed into two branches as obtained in I.2 :

$$E_\pm(k) = \frac{\hbar^2 k^2}{2m} \pm \alpha\hbar k \tag{5}$$

with $k = |\vec{k}|$. We will write $k_0 = ma/\hbar$ and $\epsilon_0 = \hbar^2 k_0^2/(2m) = m\alpha^2/2$.

II.1. We first consider the case $\alpha = 0$. Briefly recall why the density of states $D_0(\epsilon)$ of *either branch* is independent of ϵ for this two-dimensional problem. Show that $D_0(\epsilon) = mL_xL_y/(2\pi\hbar^2)$.

II.2. We return to the real problem with $\alpha > 0$ and now consider the branch $E_+(k)$.

(a) Plot the dispersion law $E_+(k)$ versus k. What is the energy range ϵ accessible in this branch ?

(b) Express k versus k_0, ϵ_0, m, $\hbar$ and the energy ϵ for this branch. Deduce the density of states :

$$D_+(\epsilon) = D_0(\epsilon)\left(1 - \sqrt{\frac{\epsilon_0}{\epsilon + \epsilon_0}}\right) \tag{6}$$

Plot $D_+(\epsilon)/D_0(\epsilon)$ versus ϵ/ϵ_0.

II.3. One is now interested by the branch $E_-(k)$.

(a) Plot $E_-(k)$ versus k and indicate the accessible energy range. How many k's are possible for a given energy ϵ ?

(b) Show that the density of states of this branch is given by

$$\epsilon < 0 \ : \ D_-(\epsilon) = 2D_0(\epsilon)\sqrt{\frac{\epsilon_0}{\epsilon + \epsilon_0}}$$

$$\epsilon > 0 \ : \ D_-(\epsilon) = D_0(\epsilon)\left(1 + \sqrt{\frac{\epsilon_0}{\epsilon + \epsilon_0}}\right) \tag{7}$$

II.4. In the rest of the problem, the temperature is taken as zero.

(a) Qualitatively explain how the Fermi energy changes with the electrons number N. Show that if N is small enough, only the branch E_- is filled.

(b) What is the value of the Fermi energy ϵ_F when the E_+ branch begins to be filled? Calculate the number of electrons N^* at this point, versus L_x, L_y, m, $\hbar$ and ϵ_0.

(c) Where are the wave vectors $\vec{k}$ corresponding to the filled states $\epsilon_F < 0$ located ?

II.5. Assume in what follows that $\epsilon_F > 0$.

(a) What are the wave vectors $\vec{k}$ for the occupied states in either branch $E_\pm$?

(b) Calculate the Fermi wave vectors k_F^+ and k_F^- versus ϵ_F, k_0, ϵ_0, m and $\hbar$.

(c) For $N > N^*$, calculate $N - N^*$ and deduce the relation :

$$N = \frac{L_x L_y m}{\pi \hbar^2} \left(2\epsilon_0 + \epsilon_F\right)$$

II.6. A weak magnetic field is now applied on the sample.

(a) Using the result of I.5, explain why the Fermi energy does not vary to first order in B.

(b) Calculate the magnetization along z of the electrons gas, to first order in B.

II.7. Now one analyzes the possibility to maintain a magnetization in the system for some time in the absence of magnetic field. At time $t = 0$, each electron is prepared in the spin state $|+\rangle_z$. The magnetization along z is thus $M(0) = N\hbar/2$. For a given electron, the evolution of $s_z(t)$ in the absence of collisions was calculated in I.4. Now this evolution is modified due to the effect of collisions. Because of the elastic collisions of an electron on the material defects, its $\vec{k}$ wave vector direction may be randomly modified. These collisions are modeled assuming that :

- they occur at regular times : $t_1 = \tau$, $t_2 = 2\tau$,..., $t_n = n\tau$,...
- the angle ϕ_n characterizing the $\vec{k}$ direction between the nth and the $(n + 1)$th collision is uniformly distributed over the range $[0, 2\pi[$ and decorrelated from the previous angles ϕ_{n-1}, ϕ_{n-2}, ..., and thus from the coefficients $a_\pm(t_n)$.
- the angles ϕ_n corresponding to different particles are not correlated.

Under these assumptions, the total magnetization after many collisions can be calculated by making an average over angles for each electron. One writes $\bar{s}(t)$ for the average of $s_z(t)$ over the angles ϕ_n.
Show that $\bar{s}(t_{n+1}) = \cos(2\omega\tau)\,\bar{s}(t_n)$.

II.8. One assumes that $\omega\tau \ll 1$.

(a) Is there a large modification of the average spin $\bar{s}$ between two collisions?

(b) One considers large times t with respect to the interval τ between two collisions. Show that the average spin $\bar{s}(t)$ exponentially decreases and express the characteristic decrease time t_d versus ω and τ.

(c) Take two materials : the first one is a good conductor (a few collisions take place per unit time), the second one is a poor conductor (many collisions occur per unit time). In which of these materials is the average spin better "protected," i.e., has a slower decrease? Comment on the result.

(d) For $\tau = 10^{-12}$ sec and $\hbar\omega=5$ μeV, estimate the spin relaxation time t_d.

II.9. To maintain a stationary magnetization in the population of N spins, one injects $g_\pm$ electrons per unit time, of respective spins $s_z = \pm\hbar/2$ into the system. For each spin value, the electrons leave the system with a probability per unit time equal to $1/t_r$. Besides, one admits that the relaxation mechanism studied in the previous question is equivalent to the exchange between the $+$ and $-$ spin populations : a $+$ spin is transformed into a $-$ spin with a probability per unit time equal to $1/t_d$, and conversely a $-$ spin is transformed into a $+$ spin with a probability per unit time equal to $1/t_d$. One calls $f_\pm(t)$ the populations of each of the spin states at time t, t_d is the spin relaxation time.

(a) Write the rate equations on $df_\pm/dt$. What are the spin populations $f_\pm^*$ in steady state?

(b) Deduce the relative spin population, that is, the quantity $P = \dfrac{f_+^* - f_-^*}{f_+^* + f_-^*}$.

In what type of material will the magnetization degrade not too fast?

Solution of the Exercises and Problems

Exercise 2000 : Electrostatic Screening

I.1. The density N/Ω and the chemical potential μ are related by

$$\frac{N}{\Omega} = \frac{1}{\Omega} \int D(\vec{k}) f_{FD}(\vec{k}) d^3\vec{k} = \frac{1}{4\pi^3} \int_0^{+\infty} \frac{4\pi k^2 dk}{\exp\beta\left(\dfrac{\hbar^2 k^2}{2m_e} - \mu\right) + 1}$$

I.2. In the high temperature and low density limit, the exponential in the denominator of the Fermi-Dirac distribution is large with respect to 1, so that

$$\frac{N}{\Omega} \simeq \frac{1}{4\pi^3} \int_0^{+\infty} e^{-\beta\left(\frac{\hbar^2 k^2}{2m_e} - \mu\right)} 4\pi k^2 dk$$

$$\frac{N}{\Omega} \simeq 2e^{\beta\mu} \left(\frac{2\pi m_e k_B T}{h^2}\right)^{3/2}$$

I.3. The total energy in presence of a potential is written

$$\varepsilon = \frac{\hbar^2 k^2}{2m_e} + V(\vec{r})$$

At high temperature and low density

$$N \simeq \frac{1}{4\pi^3} \int_\Omega e^{-\beta V(\vec{r})} d^3\vec{r} \int_0^{+\infty} 4\pi k^2 dk \, e^{-\beta\left(\frac{\hbar^2 k^2}{2m_e} - \mu\right)}$$

I.4. By definition one has

$$N = \int_\Omega n(\vec{r}) d^3\vec{r}$$

one thus identifies $n(\vec{r})$ to

$$n(\vec{r}) = n_0 \, e^{-\beta V(\vec{r})}$$

with

$$n_0 = \frac{1}{4\pi^3} \int_0^{+\infty} 4\pi k^2 dk \, e^{-\beta\left(\frac{\hbar^2 k^2}{2m_e} - \mu\right)}$$

I.5. If $\beta V(\vec{r}) \ll 1$ for any $\vec{r}$, n is uniform, i.e.,

$$n = n_0 = \frac{N}{\Omega}$$

II.1. The potential energy of an electron inside the electrostatic potential $\Phi_{\text{tot}}(\vec{r})$ is $-e\,\Phi_{\text{tot}}(\vec{r})$, the energy of a hole $+e\,\Phi_{\text{tot}}(\vec{r})$. The electron and hole densities are respectively equal to

$$n(\vec{r}) = n_0 \exp[-\beta(-e\,\Phi_{\text{tot}}(\vec{r}))]$$

$$p(\vec{r}) = p_0 \exp[-\beta(+e\,\Phi_{\text{tot}}(\vec{r}))]$$

with $n_0 = p_0 = \dfrac{N}{\Omega}$.

II.2. The total density of charges at $\vec{r}$ is equal to $e[p(\vec{r}) - n(\vec{r})]$. The charge localized at the origin also enters into the Poisson equation :

$$\Delta\Phi_{\text{tot}}(\vec{r}) + \frac{e[p(\vec{r}) - n(\vec{r})]}{\varepsilon_0 \varepsilon_r} + \frac{Q\delta(\vec{r})}{\varepsilon_0 \varepsilon_r} = 0$$

In the high temperature limit and to the lowest order

$$p(\vec{r}) - n(\vec{r}) = n_0\left[e^{-\beta e\Phi_{\text{tot}}(\vec{r})} - e^{\beta e\Phi_{\text{tot}}(\vec{r})}\right]$$

$$\simeq -2n_0\,e\beta\Phi_{\text{tot}}(\vec{r})$$

Whence the equation satisfied by $\Phi_{\text{tot}}(\vec{r})$:

$$\Delta\Phi_{\text{tot}}(\vec{r}) = \frac{2n_0\,e^2\beta}{\varepsilon_0 \varepsilon_r}\Phi_{\text{tot}}(\vec{r}) - \frac{Q}{\varepsilon_0 \varepsilon_r}\delta(\vec{r})$$

One writes $\lambda^2 = \dfrac{\Omega}{N}\dfrac{\varepsilon_0 \varepsilon_r k_B T}{2e^2}$, this is the square of a length.

II.3. If $\dfrac{N}{\Omega} = 10^{22}$ m^{-3} $T = 300$ K, $\varepsilon_r = 12.4$, then $\lambda = 30$ nm.

II.4. The effect of the charge located at the origin cannot be felt beyond a distance of the order of λ : at a larger distance $\Phi_{\text{tot}}(\vec{r})$ is rapidly constant, the electron and hole densities are constant andopposite. The mobile electrons and holes have *screened* the charge at the origin.

Exercise 2001 : Magnetic Susceptibility of a "Quasi-One-Dimensional" Conductor

I.1.

a) $\lambda = 0 \; k_B T \gg \gamma B$

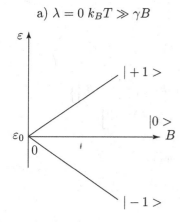

Three levels in ε_0 for $B = 0$
Comparable occupations
of the three levels

b) $\lambda \gg k_B T \gg \gamma B$

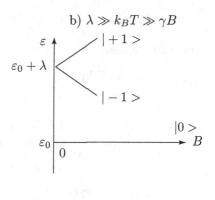

At the temperature T,
the upper levels
are weakly occupied

c) $k_B T \gg \lambda \gg \gamma B$

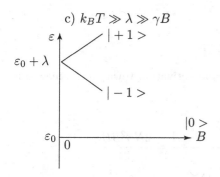

comparable occupations of the three states at T.
Little influence of λ, this case
is almost equivalent to a)

d) $\gamma B \gg \lambda \gg k_B T$

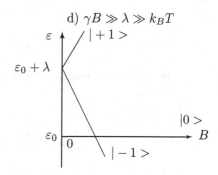

Only level $| -1 \rangle$ is populated.
The magnetization is saturated, each
ion carries a moment $+\gamma$

I.2. State $|0\rangle$, energy ε_0

$\qquad |+\rangle \qquad\qquad \varepsilon_0 + \lambda + \gamma B$

$\qquad |-\rangle \qquad\qquad \varepsilon_0 + \lambda - \gamma B$

For a single ion :

$$z = e^{-\beta\varepsilon_0} + e^{-\beta(\varepsilon_0 + \lambda + \gamma B)} + e^{-\beta(\varepsilon_0 + \lambda - \gamma B)}$$

$$= e^{-\beta\varepsilon_0}\left[1 + e^{-\beta(\lambda + \gamma B)} + e^{-\beta(\lambda - \gamma B)}\right]$$

For the whole system : $Z = (z)^N$

$$F = -k_B T \ln Z$$

$$= -Nk_B T \left\{-\beta\varepsilon_0 + \ln\left[1 + e^{-\beta(\lambda + \gamma B)} + e^{-\beta(\lambda - \gamma B)}\right]\right\}$$

$$= N\varepsilon_0 - Nk_B T \ln\left[1 + e^{-\beta(\lambda + \gamma B)} + e^{-\beta(\lambda - \gamma B)}\right]$$

$$M = -\frac{\partial F}{\partial B} = N\gamma\frac{\left[e^{-\beta\lambda + \beta\gamma B} - e^{-\beta\lambda - \beta\gamma B}\right]}{1 + e^{-\beta(\lambda + \gamma B)} + e^{-\beta(\lambda - \gamma B)}}$$

For $B \to 0$, $M \sim \dfrac{N\gamma 2\beta\gamma B e^{-\beta\lambda}}{1 + 2e^{-\beta\lambda}}$ i.e.,

$$\chi = \frac{2N\gamma^2 \beta e^{-\beta\lambda}}{\Omega(1 + 2e^{-\beta\lambda})}$$

I.3. Discussion of the limit cases :

(a) $\lambda = 0$

$$M = N\gamma\frac{e^x - e^{-x}}{1 + e^x + e^{-x}} \qquad \text{with } x = \beta\gamma B$$

In the limit $x \to 0$, one obtains $\chi = \dfrac{2N\gamma^2}{3\Omega k_B T}$ (Curie law).

(b) $\beta\lambda \gg 1 \gg \beta\gamma B = x$

$$M \simeq N\gamma e^{-\beta\lambda}(e^x - e^{-x})$$

$$\chi \simeq \frac{2N\gamma^2 e^{-\beta\lambda}}{\Omega k_B T}$$

(Curie law attenuated by the thermal activation towards the magnetic level).

(c) $1 \gg \beta\lambda \gg x$

$$M \simeq \frac{N\gamma[x - \beta\lambda + \beta\lambda + x]}{3} = \frac{2N\gamma^2 \beta B}{3}$$

$$\chi = \frac{2N\gamma^2}{3\Omega k_B T}$$

This situation is analogous to a).

(d) $\beta\gamma B \gg \beta\lambda \gg 1$

$$M \simeq N\gamma$$

The magnetization is saturated.

I.4. The high temperature behavior of χ in the figure suggests a Curie law.

The expression of χ obtained in I.2. vanishes for $\beta \to \infty$ and for $\beta \to 0$. It is positive and goes through a maximum for β of the order of $\frac{1}{\lambda}$ (exactly for $1 - \beta\lambda + 2e^{-\beta\lambda} = 0$, i.e., $\beta\lambda = 1.47$). From the temperature of the χ maximum λ can be deduced.

Note : in fact, for the system studied in the present experiment, the $\varepsilon_0 + \lambda$ level contains 3 states, and not 2 like in this exercise based on a simplified model.

II.

$$z = \sum_{-S}^{+S} e^{-l\beta\gamma B} = e^{\beta S\gamma B} \sum_{l=0}^{2S} e^{-l\beta\gamma B} = e^{\beta S\gamma B} \frac{1 - e^{-(2S+1)\beta\gamma B}}{1 - e^{-\beta\gamma B}}$$

$$f = -k_B T \ln z = -k_B T \left[S\beta\gamma B + \ln \frac{1 - e^{-(2S+1)x}}{1 - e^{-x}} \right] \quad \text{with } x = \beta\gamma B$$

$$m = -\frac{\partial f}{\partial B} = -\frac{\partial f}{\partial x}\gamma\beta = \gamma S + k_B T \frac{(2S+1)\gamma\beta e^{-(2S+1)x}}{1 - e^{-(2S+1)x}} - \frac{k_B T \beta\gamma e^{-x}}{1 - e^{-x}}$$

$$\frac{m}{\gamma} = S + \frac{2S+1}{e^{(2S+1)x} - 1} - \frac{1}{e^x - 1}$$

3 limits : $x \to +\infty$ $x \to -\infty$ $x \to 0$

(a) $x \to +\infty$ (high positive field, low temperature)

$$\frac{m}{\gamma} \to S \qquad \text{saturation}$$

(b) $x \to -\infty$ (high negative field, low temperature)

$$\frac{m}{\gamma} \to S - (2S+1) + 1 = -S \qquad \text{saturation}$$

(c) $x \to 0$: (low field, high temperature).

One expects m to be proportional to B, which allows to define a susceptibility. Therefore one has to develop m up to the term in x, which requires to express the denominators to third order.

$$\frac{m}{\gamma} \simeq S + \frac{2S+1}{(2S+1)x + (2S+1)^2\frac{x^2}{2} + (2S+1)^3\frac{x^3}{6} + \dots} - \frac{1}{x + \frac{x^2}{2} + \frac{x^3}{6} + \dots}$$

$$\frac{m}{\gamma} \simeq x\frac{S(S+1)}{3} = \frac{\gamma B S(S+1)}{3k_B T} \qquad \text{(Curie law)}$$

Exercise 2002 : Entropies of the HCℓ Molecule

1.

$$S_{3D} = Nk_B \ln\left[\frac{\Omega}{Nh^3}(2\pi mk_B T)^{3/2}\right] + \frac{5}{2}Nk_B$$

$$\mu_{3D} = -k_B T \ln\left[\frac{\Omega}{Nh^3}(2\pi mk_B T)^{3/2}\right]$$

2.

$$Z_{2D} = \frac{1}{N'!}\left(\frac{A}{h^2}2\pi mk_B T\right)^{N'}$$

$$S_{2D} = N'k_B \ln\left[\frac{A}{N'h^2}(2\pi mk_B T)\right] + 2N'k_B$$

$$\mu_{2D} = -k_B T \ln\left[\frac{A}{N'h^2}(2\pi mk_B T)\right]$$

A molecule is spontaneously adsorbed if $\mu_{2D} < \mu_{3D}$.

3. If the adsorbed film is mobile on the surface :

$$\Delta S = S_{2D} - S_{3D}$$

$$\Delta S = Nk_B \ln\left[\frac{Ah}{\Omega(2\pi mk_B T)^{1/2}}\right] - \frac{1}{2}Nk_B$$

4. For the only translation degrees of HCℓ :

$$S_{3D} = 152.9 \text{ J.K}^{-1}$$

$$S_{2D} = 48.2 \text{ J.K}^{-1}$$

The average 3D distance is of 3.3 nm, it is 0.11 nm between molecules on the surface (which is less than an interatomic distance, but one has also to account for the large corrugation of the surface).

5.

$$Z_N = \frac{1}{N!}(z_{\mathrm{rot}})^N(z_{\mathrm{tr}})^N$$

6.

$$
\begin{aligned}
z_{\mathrm{1rot}} &= \iint \frac{d\phi\, dp_\phi}{h} \exp\left(-\frac{\beta p_\phi^2}{2I_A}\right) \\
&= \frac{2\pi}{h} \int_{-\infty}^{+\infty} dp_\phi \exp\left(-\frac{\beta p_\phi^2}{2I_A}\right) \\
&= \frac{2\pi}{h} \sqrt{2\pi I_A k_B T}
\end{aligned}
$$

7.

$$
\begin{aligned}
z_{\mathrm{2rot}} &= \iiiint \frac{d\theta\, dp_\theta\, d\phi\, dp_\phi}{h^2} \exp\left(-\frac{\beta p_\theta^2}{2I}\right) \exp\left(-\frac{\beta p_\phi^2}{2I\sin^2\theta}\right) \\
&= \frac{8\pi^2}{h^2} I k_B T
\end{aligned}
$$

The expression in the course is $\dfrac{k_B T}{2hcB} = \dfrac{4\pi^2}{h^2} I k_B T$, for a symmetrical molecule, such that its physical situation is identical after a 180° rotation. But $HC\ell$ is not symmetrical

8. Rotation entropy of $HC\ell$:

$$F_{\mathrm{rot}} = -Nk_B T \ln z_{\mathrm{2rot}}$$

(the factor for indistinguishability is included inside the translation term).

$$
\begin{aligned}
S_{\mathrm{rot}} &= -\frac{\partial F_{\mathrm{rot}}}{\partial T} = Nk_B \left[\ln z_{\mathrm{2rot}} + T\frac{\partial \ln z_{\mathrm{2rot}}}{\partial T}\right] \\
S_{\mathrm{rot}} &= 33.3 \ \mathrm{J \cdot K^{-1}}
\end{aligned}
$$

Problem 2001 : Quantum Boxes and Optoelectronics

I.1. Owing to the spin degeneracies one has

$$\langle n_{E_c}\rangle = \frac{2}{e^{\beta(E_c-\mu)} + 1} \quad \text{and} \quad \langle n_{E_v}\rangle = \frac{2}{e^{\beta(E_v-\mu)} + 1}$$

I.2. At any temperature $\langle n_{E_c}\rangle + \langle n_{E_v}\rangle = 2$, in the present approximation in which the total number of electrons is 2. It follows that

$$\frac{1}{e^{\beta(E_c-\mu)}+1} + \frac{1}{e^{\beta(E_v-\mu)}+1} = 1, \quad \text{an implicit relation between } \mu \text{ and } \beta$$

I.3. This equation can be reexpressed, noting $x = e^{\beta\mu}$, $y = e^{\beta E_c}$ and $z = e^{\beta E_v}$:

$$\frac{x}{y+x} + \frac{x}{z+x} = 1$$

that is, $zx + yx + 2x^2 = zy + zx + yx + x^2$, $x^2 = yz$, whence the result $\mu = \dfrac{E_c + E_v}{2}$, independently of temperature.

I.4. One immediately deduces from I.3

$$\langle n_{E_c}\rangle = \frac{2}{e^{\beta(E_c-\mu)}+1} = \frac{2}{e^{\frac{\beta(E_c-E_v)}{2}}+1}$$

At 300 K with $E_c - E_v = 1$ eV, $\langle n_{E_c}\rangle = \dfrac{2}{e^{20}+1} = 4\times10^{-9}$.

At 300 K, at thermal equilibrium, the probability for the conduction levels to be occupied is thus extremely small.

II.1.
$$f_{nls} = \frac{1}{e^{\beta(E_{nls}-\mu)}+1}$$

As the energy E_{nls} only depends of n, consequently we will write $f_{nls} = f_n$.

II.2. Noting g_n for the degeneracy of state n, one has

$$\langle n\rangle = 1 = \sum_n g_n f_n = \sum_n 2(n+1)f_n$$

an implicit relation between μ and β.

II.3. If one had $\mu > E_c$, $2f_0$ would be larger than 1, in contradiction with $\langle n\rangle = 1$.

II.4. E_n is an increasing function of n, as is also $e^{\beta(E_n-\mu)}$, whence the result. One has $f_0 < 1/2$, and $f_{n+1} < f_n$, so that :

$$1 \geq \sum_{i=0}^{n} g_i f_i > \sum_{i=0}^{n} g_i f_n = 2f_n\sum_{i=0}^{n}(n+1) = 2f_n\frac{(n+1)(n+2)}{2}$$

$$\text{i.e., } f_n \leq \frac{1}{(n+1)(n+2)}$$

II.5. The relation obtained in II.4 implies $f_1 \leq \dfrac{1}{6}$, $f_2 \leq \dfrac{1}{12}$, etc. One then deduces that in f_1 one has $e^{\beta(E_1-\mu)} > 5$, in f_2 $e^{\beta(E_2-\mu)} > 11$, which justifies the use of the Maxwell-Boltzmann approximation, i.e.,

$$f_{nls} = \frac{1}{e^{\beta(E_{nls}-\mu)} + 1} \approx e^{-\beta(E_n-\mu)}$$

for any excited state.

Taking into account the validity of this approximation, the implicit relation between μ and β becomes

$$1 = 2f_0 + 2\sum_{n=1}^{\infty}(n+1)e^{-\beta(E_n-\mu)}$$

With $E_n = E_c + n\hbar\omega_c$, and $w = e^{-\beta(E_c-\mu)}$, this expression can be transformed into

$$\frac{1}{2} = g(\beta, w) = \frac{1}{w^{-1}+1} + w\sum_{n=1}^{\infty}(n+1)e^{-\beta n\hbar\omega_c}$$

$$= \frac{w}{w+1} + w\sum_{n=1}^{\infty}(n+1)e^{-\beta n\hbar\omega_c} \qquad (1)$$

II.6. Obviously g is a decreasing function of β at fixed w. Moreover, $\dfrac{\partial g}{\partial w} = \dfrac{1}{(w+1)^2} + \sum_{n=1}^{\infty}(n+1)e^{-\beta n\hbar\omega_c} > 0$, thus g is an increasing function of w. When T increases, β decreases. For g to remain constant [equal to $1/2$ from (1)], the fugacity w must decrease when T increases.

Since $w = e^{-\beta(E_c-\mu)}$, $\dfrac{\partial w}{\partial T} = w\dfrac{1}{kT^2}(E_c - \mu) + \beta\dfrac{\partial \mu}{\partial T}$. From II.3., $\mu < E_0 = E_c$ and the first term of the right member is > 0. Consequently, $\dfrac{\partial w}{\partial T} < 0$ implies $\dfrac{\partial \mu}{\partial T} < 0$.

II.7. From equation (1),

– when $\beta \to 0$, $\sum_{n=1}^{\infty}(n+1)e^{-\beta n\hbar\omega_c}$ tends to infinity, thus the solution w must tend to 0. Then μ tends to $-\infty$. (One finds back the classical limit at high temperature) ;

– when $T \to 0$, the sum tends to 0 and (1) becomes $\dfrac{1}{2} = \dfrac{w}{w+1}$, so that w tends to 1 in this limit : the low temperature limit of μ is then $E_0 = E_c$.

II.8. $f_0 = \dfrac{1}{w^{-1}+1}$. At low temperature, w tends to 1 and f_0 tends to $1/2$, whereas at high temperature w tends to 0 and the Maxwell-Boltzmann statistics is then applicable to the 0 level. Since w is a monotonous function of T, this latter limit must be valid beyond a critical temperature T_c, so that then $e^{\beta(E_0-\mu)} \gg 1$ $(w \ll 1)$.

II.9. The function $\chi = \displaystyle\sum_{n=1}^{\infty} e^{-\beta n \hbar \omega_c}$ is equal to $\dfrac{e^{-\beta \hbar \omega_c}}{1 - e^{-\beta \hbar \omega_c}} = \dfrac{1}{e^{\beta \hbar \omega_c} - 1}$. Its derivative

$$\frac{\partial \chi}{\partial \beta} = -\hbar \omega_c \sum_{n=1}^{\infty} n e^{-\beta n \hbar \omega_c}$$

allows to conveniently calculate the sum in the right member of (1), which is equal to :

$$\sum_{n=1}^{\infty} (n+1) e^{-\beta n \hbar \omega_c} = \chi - \frac{1}{\hbar \omega_c} \frac{\partial \chi}{\partial \beta} = \frac{1}{e^{\beta \hbar \omega_c} - 1} + \frac{e^{\beta \hbar \omega_c}}{(e^{\beta \hbar \omega_c} - 1)^2}$$

$$= \frac{2 e^{\beta \hbar \omega_c} - 1}{(e^{\beta \hbar \omega_c} - 1)^2} = -1 + \frac{e^{2\beta \hbar \omega_c}}{(e^{\beta \hbar \omega_c} - 1)^2} = -1 + \frac{1}{(1 - e^{-\beta \hbar \omega_c})^2}$$

whence the result (2).

II.10. This condition is equivalent to $w < 1/e$, and at $T = T_c$ one has $w = 1/e$. For $w = 1/e$, (2) becomes

$$1 + \frac{e(e-1)}{2(e+1)} = \frac{1}{\left[1 - \exp\left(-\dfrac{\hbar \omega_c}{k_B T_c}\right)\right]^2}$$

$$\text{i.e.,} \quad \frac{2(e+1)}{e^2 + e + 2} = \left[1 - \exp\left(-\dfrac{\hbar \omega_c}{k_B T_c}\right)\right]^2$$

which provides :

$$\frac{\hbar \omega_c}{k T_c} = -\ln\left(1 - \sqrt{\frac{2(e+1)}{e^2 + e + 2}}\right)$$

Numerically $k_B T_c = 0.65\, \hbar \omega_c$.

II.11. For $T > T_c$, the Maxwell-Boltzmann statistics is applicable to the fundamental state and consequently to any state. $f_0 = \dfrac{1}{w^{-1}+1}$ can be approached by w, which allows to write (2) as :

$$\frac{1}{2} = \frac{w}{(1 - e^{-\beta \hbar \omega_c})^2}$$

It follows that

$$f_n = e^{-\beta(E_0 + n\hbar\omega_c - \mu)} = we^{-\beta n\hbar\omega_c} = e^{-\beta n\hbar\omega_c}\frac{(1 - e^{-\beta\hbar\omega_c})^2}{2} \tag{3}$$

III.1. The average number of electrons in a state of spin s' of energy E in the valence band is also the Fermi-Dirac distribution function :

$$\langle n \rangle = \frac{1}{e^{\beta(E - \mu_v)} + 1}$$

III.2.

$$h_{n'\ell's'} = 1 - \langle n \rangle = \frac{e^{\beta(E - \mu_v)}}{e^{\beta(E - \mu_v)} + 1} = \frac{1}{e^{-\beta(E - \mu_v)} + 1} = \frac{1}{e^{\beta(-E + \mu_v)} + 1}$$

$h_{n'\ell's'}$ is indeed an expression of the Fermi-Dirac statistics, under the condition of writing $E_t = -E$. The holes chemical potential is then given by $\mu_t = -\mu_v$.

The results on the electrons in questions II.1. to II.11. apply to the holes (provided this transposition for the energy and for μ).

IV.1. If all recombinations are radiative,

$$\frac{dn}{dt} = -\frac{n}{\tau_r}$$

whence the result : $n = n_0\, e^{-t/\tau_r}$ with $n_0 = 1$.

IV.2. At $T = 0$, both f_0 and h_0 are equal to $1/2$, all the other occupation factors being equal to zero.

Consequently, one has :

$$\frac{1}{\tau_r} = \frac{f_{0,0,1}h_{0,0,1} + f_{0,0,-1}h_{0,0,-1}}{\tau_0} = \frac{2f_0h_0}{\tau_0} = \frac{1}{2\tau_0} \quad \text{i.e.,} \quad \tau_r = 2\tau_0$$

IV.3. One deduces from the figure $\tau_r = 2\tau_0 = 1$ nsec, that is, $\tau_0 = 0.5$ nsec.

IV.4. At any temperature one has

$$\frac{1}{\tau_r} = \sum_{n,l,s} \frac{f_{n,l,s}h_{n,l,s}}{\tau_0} \tag{4}$$

Moreover the average numbers of electrons and holes are equal to 1. As the occupation factors are spin-independent, for a given value of s one gets :

$$\sum_{n,l} f_n = \sum_{n',l'} h_{n'} = \frac{1}{2}$$

One deduces :

$$\sum_{n,l} f_n h_n \le \left(\sum_{n,l} f_n\right)\left(\sum_{n',l'} h_{n'}\right) = \frac{1}{4}$$

This limit being reached at $T = 0$, one then finds that at any temperature the radiative lifetime is larger than or equal to its value at $T = 0$.

IV.5. For the electrons $\hbar\omega_c \simeq 100$ meV, whereas $\hbar\omega_v \simeq 15$ meV for the holes. T_c is thus of the order of 780 K for the electrons and 117 K for the holes. Expression (2) shows that when $k_B T \ll \hbar\omega$, w is of the order of 1, which indeed constitutes the zero temperature approximation. The Maxwell-Boltzmann approximation is valid for any hole state if $T > T_c = 117$ K, from II.10.

IV.6. One thus has $f_0 = 1/2$ and all the other f_n's are equal to zero, whereas from (3),

$$h_0 = \frac{(1 - e^{-\beta\hbar\omega_v})^2}{2}$$

One then deduces that the only active transition is associated to the levels $|001\rangle$ and $|00-1\rangle$ (like at zero temperature) and that

$$\frac{1}{\tau_r} = \frac{2 f_0 h_0}{\tau_0} = \frac{h_0}{\tau_0} = \frac{(1 - e^{-\beta\hbar\omega_v})^2}{2\tau_0}$$

IV.7. This model allows one to qualitatively understand the increase of the radiative lifetime with temperature, due to the redistribution of the hole probability of presence on the excited levels. This is indeed observed above 250 K, a temperature range in which a decrease of the lifetime is observed in the experiment.

IV.8. At 200 K one approximately gets $\tau_r(T) = 2\tau_r(0)$, that is, $(1 - e^{-\beta\hbar\omega_v})^2 = \frac{1}{2}$, which gives

$$\beta\hbar\omega_v = -\ln\left(1 - \frac{1}{\sqrt{2}}\right)$$

i.e., a splitting between holes levels of $1.23\,k_B T$ for $T = 200$ K, that is, of 20 meV.

V.1. The probability per unit time of radiative recombination is $\frac{1}{\tau_r}$. This means that the variation of the average number of electrons per unit time, due to these radiative recombinations, is given by

$$\frac{d\langle n\rangle}{dt} = -\frac{\langle n\rangle}{\tau_r}$$

In presence of non radiative recombinations at the rate $\dfrac{1}{\tau_{nr}}$, the total variation becomes :

$$\frac{d\langle n\rangle}{dt} = -\frac{\langle n\rangle}{\tau_r} - \frac{\langle n\rangle}{\tau_{nr}}$$

V.2. This allows to define the total lifetime τ by

$$\frac{d\langle n\rangle}{dt} = -\frac{\langle n\rangle}{\tau_r} - \frac{\langle n\rangle}{\tau_{nr}} = -\frac{\langle n\rangle}{\tau}$$

i.e.,

$$\frac{1}{\tau} = \frac{1}{\tau_r} + \frac{1}{\tau_{nr}}$$

and a radiative yield η equal to the ratio between the number of *radiative* recombinations per unit time and the *total* number of recombinations per unit time :

$$\eta = \frac{\frac{\langle n\rangle}{\tau_r}}{\frac{\langle n\rangle}{\tau}} = \frac{\tau}{\tau_r} = \frac{\frac{1}{\tau_r}}{\frac{1}{\tau_r} + \frac{1}{\tau_{nr}}} = \frac{1}{1 + \frac{\tau_r}{\tau_{nr}}}$$

V.3. When only *radiative* recombinations are present, $\tau_{nr} \to \infty$ and the radiative yield η is equal to 1, independently of the value of the radiative lifetime. Here the yield strongly decreases above 200 K, whereas it is constant for $T < 150$ K. This implies that a nonradiative recombination mechanism becomes active for $T > 200$ K.

Correlatively, this is consistent with the decrease of the total lifetime τ in the same temperature range, whereas the results of question IV.6. show that the radiative lifetime should increase with temperature.

V.4. We found in V.2 : $\eta = \dfrac{\frac{1}{\tau_r}}{\frac{1}{\tau_r} + \frac{1}{\tau_{nr}}}$.

In this expression $\dfrac{1}{\tau_{nr}} = \gamma \displaystyle\sum_{n \geq n_0, l, s} (f_{n,l,s} + h_{n,l,s})$. At sufficiently high temperature, around 350 K, $\tau \approx 0.5$ nsec whereas $\tau_r > 2$ nsec. In this regime, the *nonradiative* recombinations determine τ, and $\eta \approx \dfrac{\tau_{nr}}{\tau_r}$. Besides, only the hole levels of index $n_0 > 0$ are occupied, they satisfy a Maxwell-Boltzmann statistics. This allows to write, using (3),

$$\frac{1}{\tau_{nr}} = \gamma \sum_{n \geq n_0} 2(n+1)e^{-\beta n \hbar \omega_v} \frac{(1 - e^{-\beta \hbar \omega_v})^2}{2}$$

Taking an orbital degeneracy equal to $n_0 + 1$ for each excited level, one has

$$\frac{1}{\tau_{nr}} = \gamma\frac{(1 - e^{-\beta\hbar\omega_v})^2}{2}2(n_0 + 1)e^{-\beta n_0\hbar\omega_v}\sum_{n\geq 0}e^{-\beta n\hbar\omega_v}$$

$$= \gamma(1 - e^{-\beta\hbar\omega_v})(n_0 + 1)e^{-\beta n_0\hbar\omega_v}$$

Using the result of IV.6,

$$\frac{1}{\eta} = \frac{\gamma(1 - e^{-\beta\hbar\omega_v})(n_0 + 1)e^{-\beta n_0\hbar\omega_v}}{\frac{(1-e^{-\beta\hbar\omega_v})^2}{2\tau_0}} = 2\gamma\tau_0\frac{(n_0 + 1)e^{-\beta n_0\hbar\omega_v}}{(1 - e^{-\beta\hbar\omega_v})}$$

$$\approx 2\gamma\tau_0(n_0 + 1)e^{-\beta n_0\hbar\omega_v}$$

In this temperature range $\ln(\eta) = \text{constant} + \dfrac{n_0\hbar\omega_v}{kT}$.

The slope of $\eta(T)$ provides an approximation for n_0 : between 200 and 350 K

$$\ln\left(\frac{\eta_{200K}}{\eta_{350K}}\right) = \ln(10) = \frac{n_0\hbar\omega_v}{k_B}\left(\frac{1}{200} - \frac{1}{350}\right)$$

which yields a value of n_0 ranging between 5 and 6.

Problem 2002 : Physical Foundations of Spintronics

Part I : Quantum Mechanics

I.1. A state having a plane wave $e^{i\vec{k}\cdot\vec{r}}$ for space part is an eigenstate of the momentum operator, with the eigenvalue $\hbar\vec{k}$. The effect of the hamiltonian on this type of state gives :

$$\hat{H}\left[\frac{e^{i\vec{k}\cdot\vec{r}}}{\sqrt{L_xL_y}}\begin{pmatrix}a_+\\a_-\end{pmatrix}\right] = \frac{e^{i\vec{k}\cdot\vec{r}}}{\sqrt{L_xL_y}}\left[\frac{\hbar^2k^2}{2m}\begin{pmatrix}a_+\\a_-\end{pmatrix} + \hbar k\alpha\begin{pmatrix}e^{-i\phi}a_-\\e^{i\phi}a_+\end{pmatrix}\right]$$

One thus sees that, if the coefficients $a_\pm$ verify the evolution equations

$$i\hbar\dot{a}_\pm = \frac{\hbar^2k^2}{2m}a_\pm + \hbar k\alpha\,e^{\mp i\phi}\,a_\mp$$

the proposed eigenvector is a solution of the Schroedinger equation $i\hbar|\dot{\psi}\rangle = \hat{H}|\psi\rangle$.

I.2.

(a) For fixed k and ϕ, one is restricted to a two-dimensional problem, only on the spin variables, and the hamiltonian is equal to

$$\hat{H}_s = \frac{\hbar^2 k^2}{2m} \mathrm{Id} + \hbar k \alpha \begin{pmatrix} 0 & e^{-i\phi} \\ e^{i\phi} & 0 \end{pmatrix}$$

where Id represents the identity matrix.

(b) It can be immediately verified that the two proposed vectors are eigenvectors of $\hat{H}_s$ with the eigenenergies $E_\pm$.

(c) The space of the one-electron states is the tensor product of the space associated to the electron motion in the rectangle $[0, L_x] \times [0, L_y]$ with the spin space. The functions $e^{i\vec{k}\cdot\vec{r}}/\sqrt{L_x L_y}$ are an orthonormal basis of the orbital space, provided that the choice of $\vec{k}$ allows this function to satisfy the periodical limit conditions : this requires $k_i = 2\pi n_i/L_i$, with n_i positive, negative or null integer. As for the two vectors $|\chi_\pm\rangle$, they constitute a basis of the spin space. Thus the vectors $|\Psi_{\vec{k},\pm}\rangle$ are an orthonormal eigenbasis of the hamiltonian $\hat{H}_0$.

I.3.

(a) The development of the initial spin state on the pair of eigenstates $|\chi_\pm\rangle$ is

$$\begin{pmatrix} a_+(0) \\ a_-(0) \end{pmatrix} = \frac{1}{2}\left(a_+(0) + a_-(0)e^{-i\phi}\right)\begin{pmatrix} 1 \\ e^{i\phi} \end{pmatrix}$$
$$+ \frac{1}{2}\left(a_+(0) - a_-(0)e^{-i\phi}\right)\begin{pmatrix} 1 \\ -e^{i\phi} \end{pmatrix}$$

The initial state (orbital $\oplus$ spin) is thus given by

$$|\Psi\rangle = \frac{a_+(0) + a_-(0)e^{i\phi}}{\sqrt{2}}\,|\Psi_{\vec{k},+}\rangle + \frac{a_+(0) - a_-(0)e^{i\phi}}{\sqrt{2}}\,|\Psi_{\vec{k},-}\rangle\,.$$

(b) One deduces that the electron spin state at time t is characterized by

$$\begin{pmatrix} a_+(t) \\ a_-(t) \end{pmatrix} = \frac{e^{-iE_+t/\hbar}}{2}\left(a_+(0) + a_-(0)e^{-i\phi}\right)\begin{pmatrix} 1 \\ e^{i\phi} \end{pmatrix}$$
$$+ \frac{e^{-iE_-t/\hbar}}{2}\left(a_+(0) - a_-(0)e^{-i\phi}\right)\begin{pmatrix} 1 \\ -e^{i\phi} \end{pmatrix}$$

After development, one finds

$$a_\pm(t) = e^{-i\hbar^2 k^2 t/2m}\left(a_\pm(0)\cos(\omega t) - i a_\mp(0)e^{\mp i\phi}\sin(\omega t)\right).$$

I.4. A simple calculation gives

$$
\begin{aligned}
s_z(t) &= \frac{\hbar}{2}\left(|a_+(t)|^2 - |a_-(t)|^2\right) \\
&= \frac{\hbar}{2}\Big[\cos(2\omega t)\left(|a_+(0)|^2 - |a_-(0)|^2\right) \\
&\qquad + 2\sin(2\omega t)\,\mathrm{Im}\left(a_+^*(0)a_-(0)e^{-i\phi}\right)\Big]
\end{aligned}
$$

One then deduces the result announced in the text.

I.5.

(a) The new term in the hamiltonian, arising from the magnetic field, commutes with the momentum operator. One can thus always look for eigenstates of the form $e^{i\vec{k}\cdot\vec{r}}\begin{pmatrix} a_+ \\ a_- \end{pmatrix}$, where $\begin{pmatrix} a_+ \\ a_- \end{pmatrix}$ is a two-component vector providing the spin state. The hamiltonian determining the evolution of this spin is written in the $|\pm\rangle_z$ basis :

$$
\hat{H}_s^{(B)} = \hat{H}_s - \frac{\hbar\gamma B}{2}\hat{\sigma}_z = \frac{\hbar^2 k^2}{2m}\mathrm{Id} + \hbar\begin{pmatrix} -\gamma B/2 & k\alpha e^{-i\phi} \\ k\alpha e^{i\phi} & \gamma B/2 \end{pmatrix}
$$

(b) One verifies that the eigenenergies of the above matrix are those given in the text.

(c) One notices that $\langle \chi_\pm|\hat{S}_z|\chi_\pm\rangle = 0$ and that $\langle \chi_\pm|\hat{\sigma}_z|\chi_\mp\rangle = 1$. One deduces the formulae given in the text.

(d) For a large B, the eigenstates of $\hat{H}_s^{(B)}$ are practically equal to the eigenstates $|\pm\rangle_z$ of $\hat{S}_z$. More precisely, for $\gamma < 0$ (electron), one has $|\chi_\pm^{(B)}\rangle \simeq \pm|\pm\rangle_z$. Consequently, $\langle \chi_\pm^{(B)}|\hat{S}_z|\chi_\pm^{(B)}\rangle \simeq \pm\hbar/2$.

Part II : Statistical Physics

II.1. For $\alpha = 0$, one has $E_\pm = \hbar^2 k^2/(2m)$. The number of quantum states in a given spin state, with a wave vector of coordinates in the range (k_x to $k_x + dk_x$; k_y to $k_y + dk_y$) is equal to

$$
d^2\mathcal{N} = \frac{L_x L_y}{(2\pi)^2}dk_x dk_y = \frac{L_x L_y}{(2\pi)^2}k\,dk\,d\phi
$$

i.e., by integration over the angle ϕ :

$$
d\mathcal{N} = \frac{L_x L_y}{2\pi}k\,dk = \frac{L_x L_y}{2\pi}\frac{m}{\hbar^2}d\epsilon
$$

The density of states is thus independent of the energy ϵ and equal to $D_0(\epsilon) = L_x L_y m/(2\pi\hbar^2)$.

II.2.

(a) The dispersion law in this branch is written

$$E_+(k) = \frac{\hbar^2(k + k_0)^2}{2m} - \epsilon_0$$

This corresponds to a portion of parabola, plotted in Fig.1. The spectrum of the allowed energies is $[0, \infty[$.

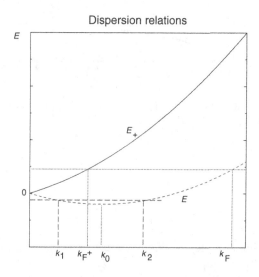

FIG. 1 : Typical behavior of the dispersion laws $E_\pm(k)$ in either branch.

(b) For the branch $\epsilon = E_+(k)$,

$$k = -k_0 + \sqrt{\frac{2m}{\hbar^2}(\epsilon + \epsilon_0)}$$

and thus

$$D_+(\epsilon) = \frac{L_x L_y}{2\pi} k \frac{dk}{d\epsilon} = D_0(\epsilon)\left(1 - \sqrt{\frac{\epsilon_0}{\epsilon + \epsilon_0}}\right)$$

The corresponding curve is plotted in 2.

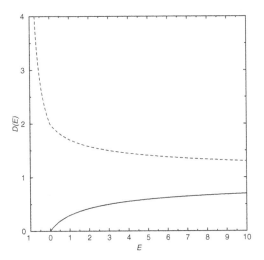

FIG. 2: Densities of states D_+/D_0 (lower curve) and D_-/D_0 (upper curve), plotted versus ϵ/ϵ_0.

II.3.

(a) For the branch $\epsilon = E_-(k)$,

$$E_-(k) = \frac{\hbar^2(k - k_0)^2}{2m} - \epsilon_0$$

the variation of which is given in Fig.1. The spectrum of the permitted energies is $[-\epsilon_0, \infty]$.

– For $-\epsilon_0 < \epsilon \leq 0$, there are two possible wavevector moduli k for a given energy ϵ :

$$k_{1,2} = k_0 \pm \sqrt{\frac{2m}{\hbar^2}(\epsilon + \epsilon_0)}$$

– For $\epsilon > 0$, there is a single allowed wavevector modulus :

$$k = k_0 + \sqrt{\frac{2m}{\hbar^2}(\epsilon + \epsilon_0)}$$

(b) One then deduces, for $\epsilon > 0$

$$D_-(\epsilon) = D_0(\epsilon)\left(1 + \sqrt{\frac{\epsilon_0}{\epsilon + \epsilon_0}}\right)$$

and for $\epsilon < 0$ (adding both contributions corresponding for either sign in the $k_{1,2}$ formula) :

$$D_-(\epsilon) = 2D_0(\epsilon)\sqrt{\frac{\epsilon_0}{\epsilon + \epsilon_0}}$$

The corresponding curve is plotted in Fig. 2.

II.4.

(a) Assume that the electrons are accommodated one next to the other in the available energy levels. Since the temperature is taken to be zero, the first electron must be located on the fundamental level, the second one on the first excited level, and so forth. Thus the first levels to be filled are those of the E_- branch, with k around k_0, the Fermi energy ϵ_F being a little larger than $-\epsilon_0$ for N small. When N increases, the energy ϵ_F also increases. It reaches 0 when all the states of the E_- branch, of wave vectors ranging between 0 and $2k_0$, are filled. For larger numbers of electrons, both branches $E_\pm$ are filled.

(b) The E_+ branch remains empty until ϵ_F reaches the value $\epsilon_F = 0$. There are then

$$N^* = \int_{-\epsilon_0}^{0} D_-(\epsilon)d\epsilon = \frac{2L_xL_ym}{\pi\hbar^2}\epsilon_0$$

electrons on the E_- branch.

(c) In the E_- branch, the filled states verify $E_-(k) < \epsilon_F$, i.e., $k_1 < k < k_2$ with :

$$k_{1,2} = k_0 \pm \sqrt{\frac{2m}{\hbar^2}(\epsilon_F + \epsilon_0)}$$

(see the dashed lines in Fig.1). The extremities of the corresponding $\vec{k}$ vectors are thus located inside a ring, limited by two Fermi "surfaces" consisting of the circles of respective radii k_1 and k_2.

II.5.

(a) When N is larger than N^*, one has to add the extra $N - N^*$ electrons in both branches, up to the Fermi level. The situation is then given by the dotted line in Fig.1. For either $E_\pm$ branch, the extremities of the wave vectors $\vec{k}$ corresponding to the filled states are located inside the disk defined by $E_\pm(k) < \epsilon_F$.

(b) For the E_+ (resp. E_-) branch, the radius of the disk corresponding to the occupied states is determined by $E_+(k_F^+) = \epsilon_F$ (resp. $E_-(k_F^-) = \epsilon_F$), with :

$$k_F^\pm = \mp k_0 + \sqrt{\frac{2m}{\hbar^2}(\epsilon_F + \epsilon_0)}$$

In either branch, the Fermi "surface" is a circle for this two-dimensional problem.

(c) One finds :

$$N - N^* = \int_0^{\epsilon_F} (D_+(\epsilon) + D_-(\epsilon))\, d\epsilon = \frac{L_xL_ym}{\pi\hbar^2}\epsilon_F$$

One then deduces the relation between the electrons number and the Fermi energy :

$$N = \frac{L_x L_y m}{\pi \hbar^2} (2\epsilon_0 + \epsilon_F)$$

II.6.

(a) The energies $E_\pm^{(B)}$ in either branch only vary to second order in B, i.e., one has $E_\pm^{(B)} = E_\pm^0 + O(B^2)$. This implies that the Fermi energy too only varies to second order in B.

(b) The total magnetization at zero temperature is the sum of the average values of $\hat{S}_z$ in each filled quantum state, i.e.,

$$\begin{aligned}
M &= \sum_{k < k_F^+} \langle \Psi_{\vec{k},+}^{(B)} | \hat{S}_z | \Psi_{\vec{k},+}^{(B)} \rangle + \sum_{k < k_F^-} \langle \Psi_{\vec{k},-}^{(B)} | \hat{S}_z | \Psi_{\vec{k},-}^{(B)} \rangle \\
&= - \int_0^{k_F^+} \frac{\hbar \gamma B}{4 \alpha k} \frac{L_x L_y}{2\pi} k dk + \int_0^{k_F^-} \frac{\hbar \gamma B}{4 \alpha k} \frac{L_x L_y}{2\pi} k dk
\end{aligned}$$

One thus finds :

$$M = \frac{\hbar \gamma B}{4\alpha} \frac{L_x L_y}{2\pi} (k_F^- - k_F^+) = \frac{L_x L_y}{4\pi} \gamma B m$$

II.7. One returns to the expression of question I.4, replacing $a_\pm(0)$ by $a_\pm(t_n)$ and $a_\pm(t)$ by $a_\pm(t_{n+1})$. One averages over the angle ϕ_n, using the fact that this angle is not correlated to $a_\pm(t_n)$: $\overline{a_+^*(t_n) a_-(t_n) e^{-i\phi_n}} = 0$. One thus finds that the term in $\sin(2\omega\tau)$ does not contribute and then gets the relation given in the text : $\bar{s}(t_{n+1}) = \cos(2\omega\tau) \bar{s}(t_n)$.

II.8.

(a) If $\omega\tau \ll 1$, then $\cos(2\omega\tau) \simeq 1 - 2\omega^2\tau^2$. The change of $\bar{s}(t)$ in a time interval τ between two collisions is then weak.

(b) In the time interval $t \gg \tau$, the number of collisions $n = t/\tau \gg 1$ takes place (the fact that t is not an exact multiple of τ plays a negligible role). Consequently,

$$\bar{s}(t) = \bar{s}(0) \, (\cos(2\omega\tau))^n = \bar{s}(0) \, \exp\left(n \ln(\cos(2\omega\tau))\right)$$

Using $\ln(\cos(2\omega\tau)) \simeq \ln(1 - 2\omega^2\tau^2) \simeq -2\omega^2\tau^2$, one obtains $\bar{s}(t) = \bar{s}(0) \, e^{-t/t_d}$ with $t_d = 1/(2\omega^2\tau)$.

(c) The shorter τ, the longer t_d. Paradoxically, a poor conductor, in which the time interval between collisions is very short, leads to a longer spin relaxation time than a good conductor. This surprising result can be put together with the quantum Zenon effect, in which one finds that a frequent observation of a system prevents its evolution.

(d) One finds $\omega = 7.6 \times 10^9$ sec^{-1} and $t_d = 8.7 \times 10^{-9}$ sec. As announced in the introduction of the text, the spin relaxation time t_d can be much longer than the interval τ between collisions, corresponding to the relaxation time of the electron velocity.

II.9. Let us write the rate equation for the spins $+$ and $-$ populations :

$$\frac{df_+}{dt} = g_+ - \frac{f_+}{t_r} - \frac{f_+}{t_d} + \frac{f_-}{t_d} \qquad \frac{df_-}{dt} = g_- - \frac{f_-}{t_r} - \frac{f_-}{t_d} + \frac{f_+}{t_d}$$

In steady-state regime, one has $df_+/dt = df_-/dt = 0$, which leads to

$$f_+^* = \frac{\left(\dfrac{1}{t_r} + \dfrac{1}{t_d}\right)g_+ + \dfrac{1}{t_d}g_-}{\dfrac{1}{t_r^2} + \dfrac{2}{t_r t_d}} \qquad f_-^* = \frac{\left(\dfrac{1}{t_r} + \dfrac{1}{t_d}\right)g_- + \dfrac{1}{t_d}g_+}{\dfrac{1}{t_r^2} + \dfrac{2}{t_r t_d}}$$

and thus

$$P = \frac{g_+ - g_-}{g_+ + g_-} \frac{1}{1 + 2\dfrac{t_r}{t_d}}$$

To maintain a high polarization, at fixed injection rate, one must minimize the denominator, and thus select a material in which t_d is as large as possible, consequently in which collisions occur very frequently. Examples of these developments, together with references, may be found in the paper in Physics Today volume **52** (1999) p.33.

Index